AF539597

INNOVATIVE ENAMELING

Louie S. Taylor

VANTAGE PRESS
New York / Washington / Atlanta
Los Angeles / Chicago

FIRST EDITION

Published by Vantage Press, Inc.
516 West 34th Street, New York, New York 10001

Manufactured in the United States of America
ISBN: 0-533-06591-7

Library of Congress Catalog Card No.: 85-90065

This book is dedicated to
my dear departed wife,
MARIE ELEANOR TAYLOR,
with this poem,
composed on Mother's Day 1984

My dear sweet wife has gone to rest,
Still we're not too far apart;
A message from God, who knows her best,
Says, "Her heaven is in my heart."

Contents

Preface

Enameling on metal is one of the older forms of handicraft, dating back to the sixth century B.C. Today it is an appealing leisure-time activity, offering an opportunity for self-expression and using the imagination.

The author spent in excess of thirty years experimenting with the various phases of enameling. Many hours were devoted to research in his attempt to learn about the special characteristics and physical properties of the many kinds and colors of enamels. During this period, teaching enameling to adult students became a major effort. This did, however, provide an opportunity to explore and evaluate any new discoveries. During these fruitful years, many new and innovative ideas and techniques were developed.

Each of the articles shown in the photographs in this book was created and constructed by the author. Since he has been working with all kinds of metals for more than sixty years, it is inevitable that there be an accumulation of treasures.

In this book, the author makes no attempt to dwell on the traditional techniques or the historical aspects of enameling. As interesting as these phases may be, there is no need to present material that is already covered in many other books.

With these thoughts in mind, the author has divided the book into a series of chapters discussing the more interesting and gratifying ways of enameling. The book is written in a simple, step-by-step, easily understood manner, each chapter relating to one specific feature of enameling.

The Joy of Sharing

There are many ways of expressing caring,
The one I prefer is *the joy of sharing.*

Any talents I possess were given to me,
And I'm so grateful such things are free.

So why should I keep them to myself,
Hidden away on a conceited shelf?

There may be others with the desire
To find an activity in which to aspire.

I receive my reward when I succeed
In arranging my thoughts so that others may read.

So I go on writing, with simple pride,
And share with others the things I've tried.

CHAPTER ONE

Materials, Tools, and Equipment

Base Metals

Copper, gold, silver, silver-plated steel, and aluminum can all be successfully enameled. Each, however, requires special treatment.

Copper is the metal most often used by the hobbyist as well as by most professional enamelists. It is easy to use, inexpensive, and can be purchased in a variety of precut shapes. Eighteen gauge is recommended, either precut or in sheets from which the various shapes may be cut. Unless otherwise indicated, the chapters on the following pages will be concerned primarily with copper as a base metal.

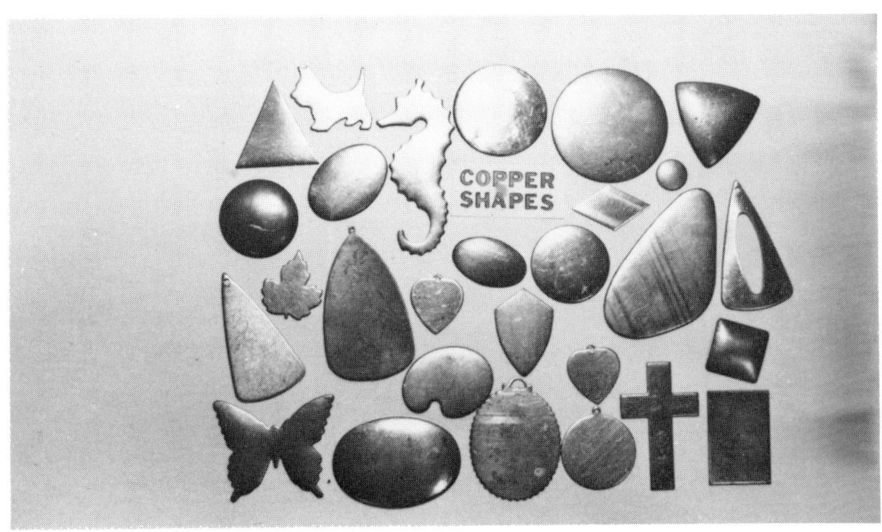

Assorted precut, 18-gauge copper shapes

Enamels

Enamels may be purchased at most hobby shops and can be obtained in one- or two-ounce packages or in larger quantities. The price is approximately fifty cents per ounce in small quantities and four dollars per pound in larger quantities, certain colors being more expensive than others. For instance, reds, oranges, pinks, and a few others contain gold oxide. This accounts for the additional cost of these colors.

A starter set should contain about ten or twelve colors, predominantly opaque colors with a few transparents. Each of the colors should be put into a two-ounce, wide-mouth jar. Two layers of nylon hosiery should be stretched over the opening and secured with a rubber band. The nylon serves as a sifter when the enamel is dusted from the jar to the copper and provides a convenient sieve that assists in removing any contaminating particles when the enamel is returned to the jar.

In order to have a quick and easy identification system, the beginner should provide each jar with a sample piece of copper that has been enameled and glued to the top of the lid. This is best accomplished by using a half-inch circle of copper for each of the opaque colors, and a half-inch square of copper for each of the transparent colors. The circle can be thought of as signifying an "O" for opaque, and the square a window for transparent. This procedure provides a good learning experience, acquainting the beginner with the characteristics of each color and with the true appearance of each enamel after it has been properly fired. This exercise should be accomplished at the outset, before the actual enameling is started. Each color should then be labeled, on the jar, with the name of the color and number. This makes it possible to have ready reference in case a new supply is required.

Enameling Kits

Beginners may prefer to purchase enameling kits. These are available in craft supply stores and are reasonably priced but somewhat limited in their usefulness. The electric kiln that is

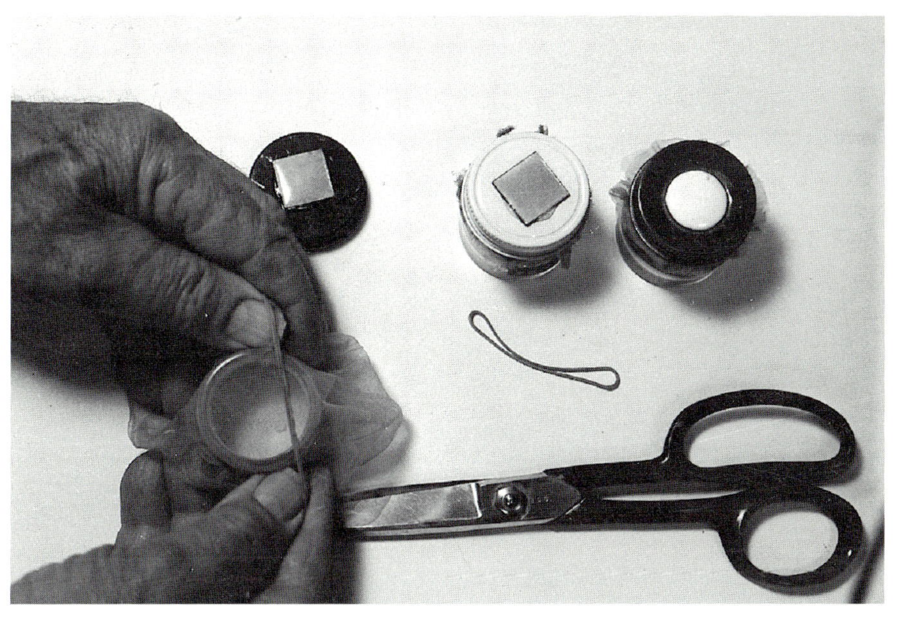

Preparing the jars for dusting the enamels

Proper identification of the enamel jars

included is about five inches in diameter and will serve well for firing most small pieces. The kit usually contains several ready-cut shapes of copper, an inexpensive spatula, about ten vials (each holding about half an ounce of enamel), a bottle of 7001 enameling solution, and a swirling wire.

Binders

A binder is a solution used to keep the powdered enamel in place when it is being fired. It is applied with a brush, or it can be sprayed over the surface of the metal. The enamel is then sifted over the binder, causing the enamel to adhere to the surface. Any binder solution that is used must burn out, leaving no residue or discoloration.

1. *Scalex, used to help prevent firescale*
2. *Ribbon solder, for soft soldering*
3. *Sparex No. 2. Mixed with water, this becomes pickling solution.*
4. *Soldering paste or soldering flux*
5. *Agar solution, one of the binders*
6. *7001 enameling solution or binder*
7. *Klyr-Fire, another binder*

There are several solutions that can be used as binders. Gum tragacanth is one of these. It tends to dry very rapidly, and this is a disadvantage in most cases. Klyr-Fire is another type of binder. It has many uses and they will be explained in detail in another chapter. Usually the most satisfactory binder is a solution called Formula 7001. More information concerning it and other binders will be covered in the following chapters as the need arises.

Pickling Materials

Any exposed metal surface that is not sufficiently protected or covered with enamel will oxidize when fired, and an unsightly fire-scale will be produced. The encrustation can be removed by placing the copper in a pickling solution. Sometimes diluted nitric or sulphuric acid is used for this purpose, but it is recommended that the student use a commercial product that is sold under the trade name of Sparex No. 2. It is cleaner, safer, and superior in many respects. It is manufactured in the form of a dry, granular acid compound. Sparex No. 2 may be purchased at the hobby shop and combined with water according to the directions on the can. A Pyrex dish serves as an excellent receptacle in which to prepare it. Avoid breathing the fumes that arise from the solution while copper pieces are being pickled. The pieces should be lifted out of the solution with a spatula, not the fingers.

Protectors

A commercial product called Scalex is manufactured for the purpose of preventing fire-scale. It is a thick liquid-clay product that can be painted on the back of the copper piece before any enameling is done. It is dried under a heat lamp or in some warm place, or it can be dried overnight at room temperature. The front or other side is then enameled in the usual way, while the dry Scalex coating is retained on the back. When the article is fired, the Scalex becomes a brittle scale that flakes off, leaving a clean surface. This process must be repeated each time a new coat of enamel is applied. Care must be exercised that no Scalex particles get into the enamel. It is seldom necessary to coat small pieces with Scalex, as they can be cleaned easily by other methods.

Fine-Line-Black

Fine-line-black is a product used for making lines on enameled pieces. It is a mixture of a special oil and finely ground glass. Fine-line-black is applied with a small brush or a pen. The lines are placed on the work to either outline or embellish the design. After the lines are drawn, the work must have time to dry before firing. Any unwanted black or errors can be rectified by scraping with a sharp instrument. Be certain that the fine-line-black is completely dry; otherwise it will adhere to the tool and become irregular and rough. When everything is satisfactory, the piece is fired at about 1300°F.

Tools

There are a few small tools that are essential in order to do enameling with ease and success. Many of these tools will be found in the average home workshop or can be made by the craftsman or purchased at little expense. The following list will serve to acquaint the beginner with those that are most needed. The tools shown in the accompanying photograph are numbered to correspond to those in the list.

1. tin snips or metal shears
2. ball-peen hammer
3. fret or jeweler's saw
4. flat-nose pliers
5. nippers
6. round-nose pliers
7. copper-pointed swirling rod
8. stainless-steel spatula
9. stainless-steel swirling rod
10. palette knife
11. tweezers
12. small scissors
13. copper spatula
14. rawhide mallet
15. small mill file

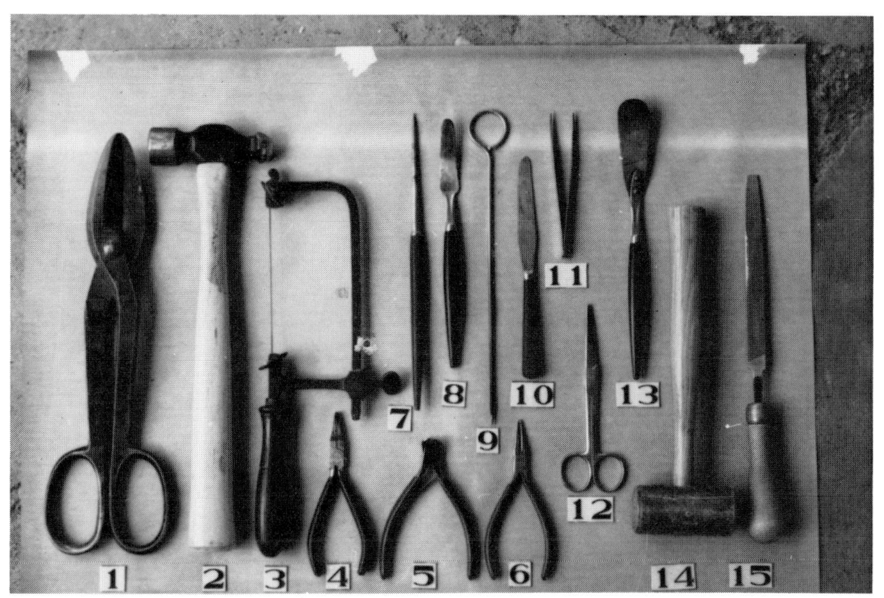

Essential tools for starter set

Occasionally, special tools and pieces of equipment are required to accomplish some of the less common techniques. When these particular tools are unavailable, they have to be made. The necessary information will be found in the appropriate chapter.

Equipment

Kilns

There are many kinds of kilns available to the enamelist. The amateur will probably be interested in obtaining one of the less expensive varieties, the kind that uses regular house current. The kilns shown in the photographs are all suitable for firing the average piece of enamel, and they all work on 115 volts and plug into the ordinary convenience outlet. Some of the larger and more expensive kilns are provided with a pyrometer to regulate the

115-volt electric kilns (on this page and facing page)

temperature and to indicate when the proper heat has been reached. The price of kilns ranges from approximately $100 to as little as $10. The elements in most kilns are replaceable; when one burns out or becomes damaged, it can be taken out and a new one installed.

Sometimes after the kiln has undergone long usage, enamel accumulates on the hearth of the kiln, and when it is heated it becomes sticky. This condition can be rectified by coating the hearth with a material known as Kiln Wash. It is a dry clay powder obtainable in most hobby shops. By mixing it with water a thick paste can be made. The paste is brushed over the hearth to a smooth, even coat. It will dry in about an hour, and then the kiln will be ready for use. This application is a permanent coating that covers the troublesome sticky places and refinishes the surface. This procedure may be repeated when the need arises.

Trivets and Stilts

A trivet is a device that is used to support an object when it is being fired in the kiln. When both sides of the work are enameled, it is necessary to use a trivet in order to prevent the

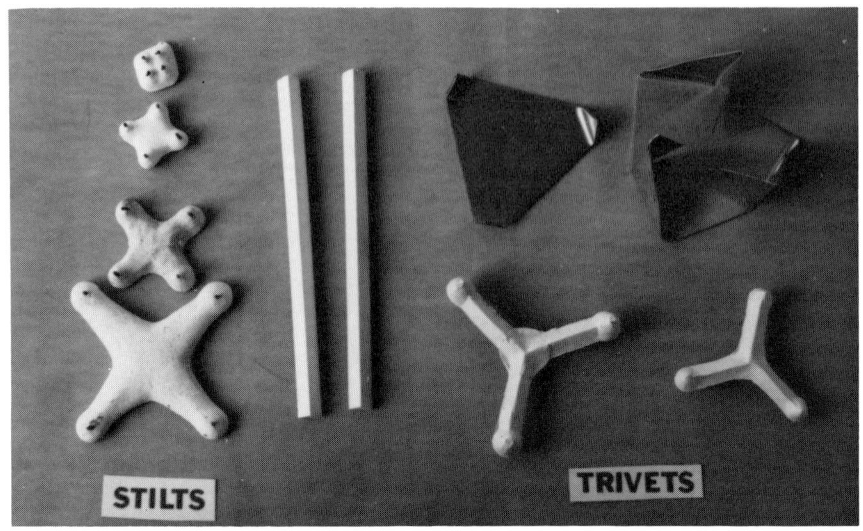

Trivets and stilts

enamel from touching the hearth of the kiln. The piece to be fired is placed in such a manner that only the edge comes in direct contact with the trivet, or it is placed on top of the trivet. If the latter method is used, the enamel on the bottom of the work will stick to the points of the trivet, causing small indentations. These indentations are not usually objectionable, and they can easily be removed by smoothing them down with a file or hidden by covering them with small felt pads.

Trivets are made in many styles and are usually constructed of stainless steel, a metal that does not scale off when heated.

Stilts are similar to trivets, though the former are commonly made of ceramic material and as a rule they are designed with four steel points. They are obtainable in many sizes and a glance through a supplier's catalog will give the student additional information concerning them.

Firing Rack

A firing rack made of heat-resistant nichrome wire is used in the larger kilns.

The piece to be fired is set on the rack and the rack is then placed in the kiln. Be certain that no part of the trivet touches the electrical element. This may cause a short and render the kiln useless until a new element is installed.

Firing Fork

A firing fork is necessary in order to transfer the rack to and from the kiln. The long handle is provided to prevent you from getting burned. Do not allow the tips of the fork to touch the electrical elements.

Torches

A torch is an essential piece of equipment. There are several kinds of torches, but the type that burns acetylene gas from a tank, and oxygen from the air, will be found to be the most satisfactory. The key or wrench on the tank valve is to be left in place at all times. There may be a sudden need to turn the tank off to avoid a fire or leakage. This is very important.

Firing rack, firing fork, and spatula

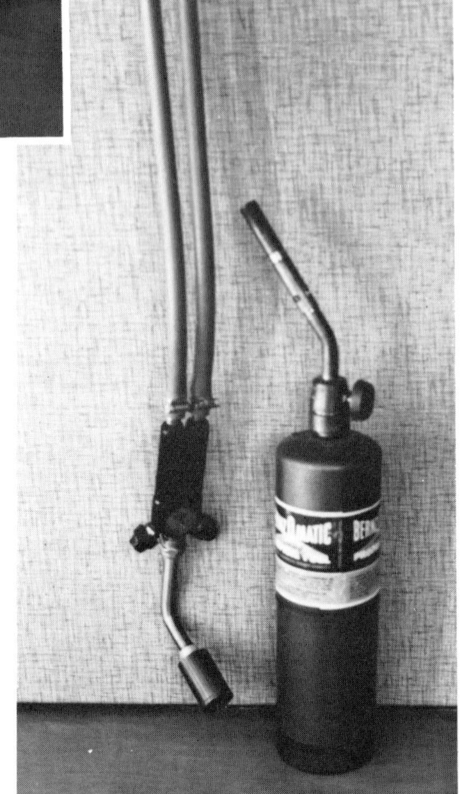

Propane torch and natural-gas torch

The bottled-propane torch is good for small pieces. It should be held firmly in the hand or secured in some way to prevent it from rolling or becoming out of control. A severe burn could result from an insecure arrangement of the torch. The round tank should be strapped to a flat board that is large enough to prevent accidental movement.

A special torch used by glass blowers is a convenient piece of equipment and serves well on special occasions. It uses natural gas and compressed air. When adjusted properly, the flame can be applied directly to the enameled surface without discoloring or staining the enamel, this being especially advantageous when transparent enamels are used. This equipment is more sophisticated and less likely to be available. It requires an air compressor and natural gas piped to the location where it is to be used. This is not recommended for the beginner; however, it is a subject that should be mentioned.

Metal Grid

A special grid will be necessary when using a torch for firing enameled pieces. The student can have it made at a welding shop or may be able to construct it himself. It consists of a framework of ¼-inch steel rod, covered with black metal lath. The metal lath is stretched over the frame and its overlapping edges are bent around the frame to hold it securely in place.

To make the frame, a piece of ¼-inch round steel welding rod 36 inches long is required. Thirteen and one-half inches from each end, the rod is heated and bent at right angles, leaving a 9-inch section of rod between the two corners. Another 9-inch piece of rod is welded between the two 13½-inch members, forming a 9-inch square with two 4½-inch legs. (These legs will keep the grid a distance away from the bench.) The metal lath is turned over the edges, forming the grid. The work to be fired is placed on the grid in a convenient way and fired from the underside.

Before the grid is used, it will be necessary to drill two holes into the bench to accommodate the two legs. This leaves the grid projecting outward from the work bench in an excellent position for firing with a torch.

"B" tank of acetylene

Acetylene torch, grid,
and water container

Framework for grid
constructed of ¼" mild steel

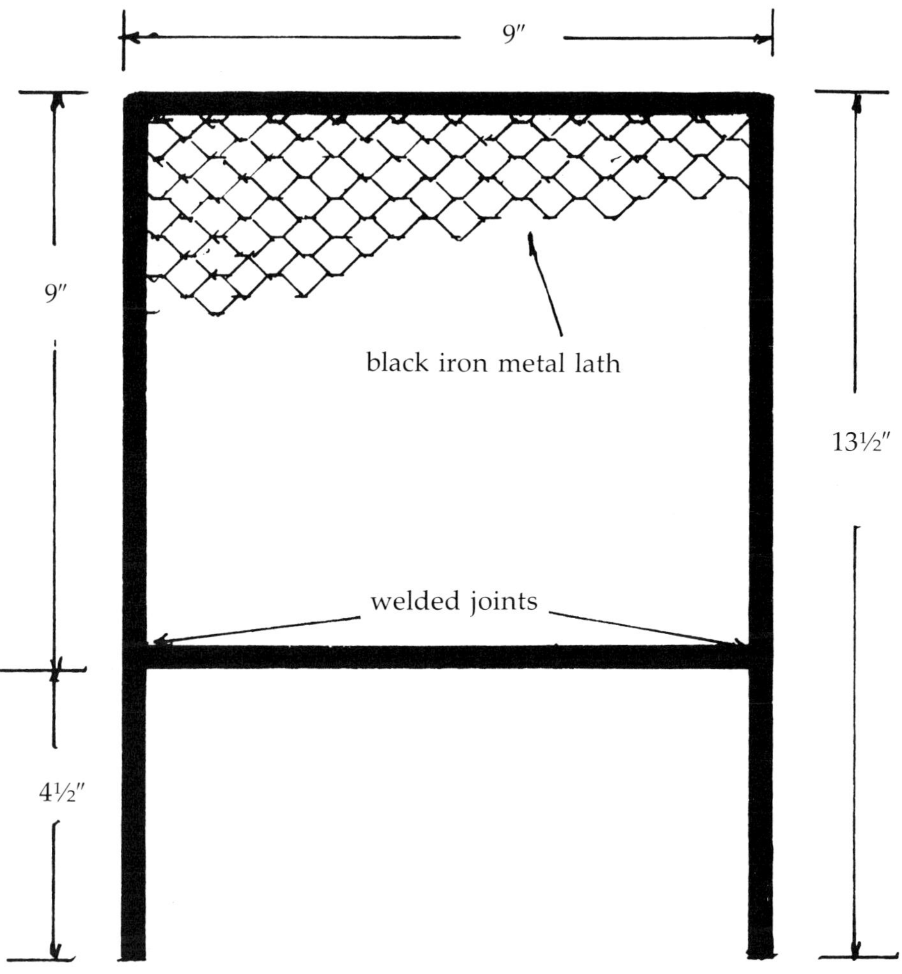

CHAPTER TWO

Characteristics of Enamels

The enamel that is used by the enamelist is primarily glass, technically called *silica*. To provide special properties such as luster, fusability, opacity, and elasticity, small quantities of potash (potassium carbonate), borax (sodium tetraborate), and soda (sodium carbonate) are added.

Each individual color is created by the addition of a certain oxide of metal. The reds and oranges are produced by adding gold oxide, and this accounts for the fact that they are more expensive than other colors. The green and turquoise colors are produced by adding an oxide of copper. The blues, for the most part, are made by adding an oxide of cobalt; yellow, uranium oxide; black, iridium oxide; white, tin oxide; purple, manganese oxide; and the grays, platinum oxide.

When a batch of molten enamel is removed from the retort or furnace during the manufacturing process, it is cooled and broken into small pieces. This raw material is called frit and is made in approximately two hundred colors. Frit is ground to various grades of powder or left in small fragments referred to as lumps.

Kinds of Enamels

Fluxes

Enamels that contain no metallic oxides are colorless, transparent enamels and are known as fluxes. There are several kinds of fluxes, each melting at a different temperature and each serving

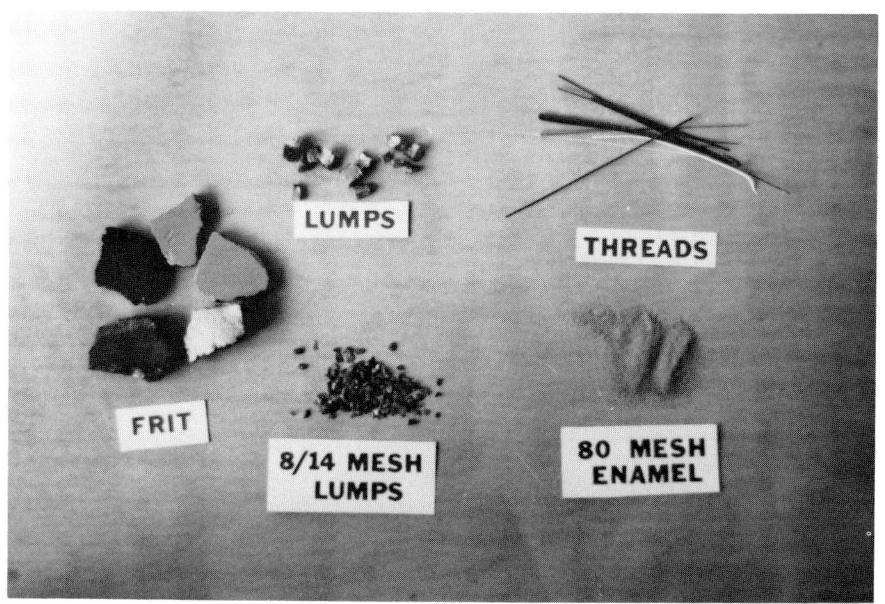

Kinds of enamels

a different purpose. Clear flux when used over any metal will expose the color of the metal. When clear flux is fired over copper, the copper will show through the flux. Frequently flux is used under transparent enamels to enhance their colors.

Each base metal requires its own special flux. There is one manufactured especially for copper, one for silver, another for silver-plated steel, and so forth. These differ from soldering fluxes.

Opalescent Enamels

Opalescent enamels, sometimes referred to as translucent enamels, are semitransparent. The choice of color is limited, but when a compromise between a transparent and an opaque is desired, they often serve a definite purpose.

Crackle Enamels

Crackle enamels are sometimes called slush. They produce a crackled effect when fired over a prefired undercoat. More information concerning crackle enamel will be found in another chapter.

Painting Colors

Still other enamels are known as painting colors. They consist of a limited number of finely ground powders that are mixed with a special oil and used as paint. A small quantity of powder and oil is placed on a glass slab and with a palette knife they are blended to a smooth consistency. Intermediate shades can be obtained by mixing two or more colors. This feature applies to painting colors only.

Opaque and Transparent Enamels

The enamels most commonly used are opaques and transparents. They are usually ground to 80 mesh and are obtainable in a variety of colors. As indicated by the name, opaques are solid colors that completely conceal the undersurface, whereas transparents permit reflection of the light and somewhat reveal the color of the copper beneath.

The firing temperature of the individual colors varies from 1450 to 1550°F. The enamels that melt in the lower temperature range are termed soft, or low firing. Those in the middle group are termed medium fusing or medium firing, and those in the upper range are referred to as hard, or high firing.

Every color has its own special characteristics and peculiarities that must be considered. The opaque reds are usually found to be in the low-firing group, and for that reason the danger of overfiring is greater. If overfired they become dull and dark, sometimes even black.

Most of the opaque greens and some of the blues belong to the low-firing group and are subject to color changes with overfiring. They tend to become transparent and somewhat darker. The yellows and blacks are quite stable, usually maintaining their true colors and generally not subject to the problems of overfiring.

Overfiring often causes the base metal to oxidize, and the oxides to penetrate the enamel. This penetration is referred to as "bleeding," and it occurs most frequently when the enamel is not thick enough. For example, if a thin coat of opaque white is applied to a piece of copper and the piece is overfired, the copper will oxidize and an undesirable green stain will appear. This can usually be remedied by an additional coat of white and a subsequent refiring, or it can be prevented in the first place by an application of a heavier coat, of any color.

In order to be properly fused, transparent enamels require slightly more heat than opaque enamels. Transparents are less likely to burn at the higher temperatures.

The most brilliant transparent colors are achieved when applied over white or flux.

Some enameled pieces, especially black pieces, lose their luster and become dull when allowed to remain in the pickling solution for too long.

Easy Scroll enamels melt at a lower temperature than the more common enamels. They have a peculiar characteristic of penetrating into the undercoat of any opaque enamel that has been used. For example, if an opaque undercoat is dusted on a piece of copper, and a few lumps of transparent amber #728 are placed at random on top of the same piece fired, the lumps will melt and penetrate the undercoat. The finished color will differ somewhat, but the final results will be pleasing and interesting.

CHAPTER THREE

Fundamentals of Enameling

A copper shape is selected. It is cleaned on the upper side with fine steel wool to remove all grease, dirt, and fingerprints. The cleaning is done in an area that is removed from where the enameling is to take place; otherwise small fragments of steel wool may contaminate the enamel.

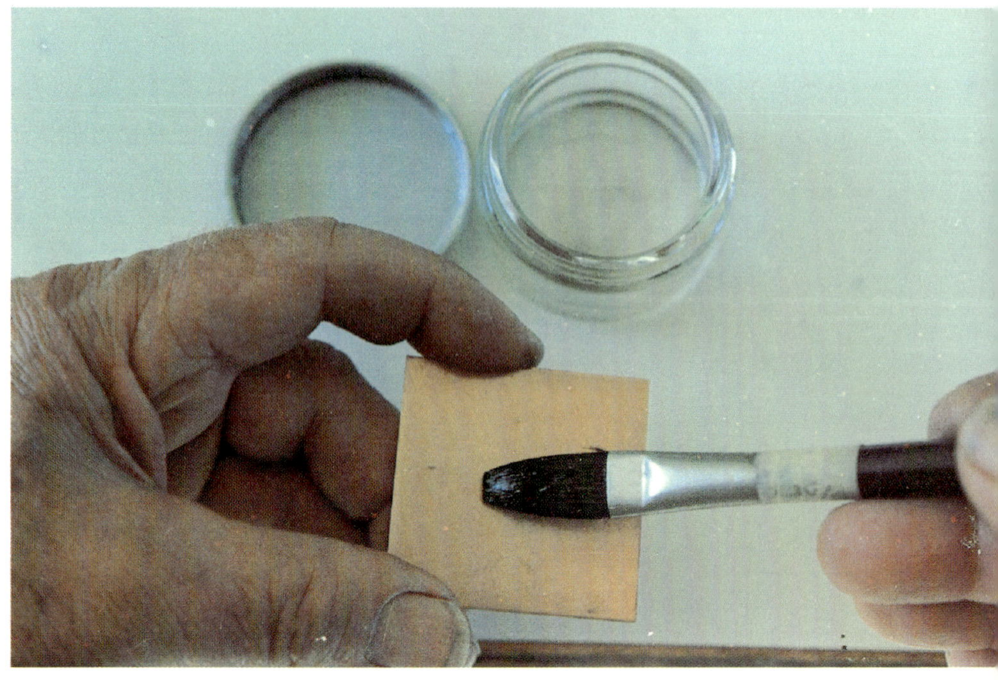

Applying 7001 enameling solution to copper piece

The student holds a copper piece in one hand and with a brush applies a thin coating of binder. Gum tragacanth may be used, but Formula 7001 enameling solution is recommended. The student is cautioned against using too much of the solution because it may cause blisters in the enamel during firing.

The piece of copper is placed on a clean, stiff sheet of paper and a layer of powdered enamel is sprinkled over it. The enamel must completely cover the copper and be evenly distributed. Special care must be taken to dust the enamel out to the edges of the piece of copper. Insufficient enamel at the edges will result in an unsightly border. It is rather difficult to judge the amount of enamel that is needed. A good rule to follow is to use enough enamel to generously cover the metal, being certain that there are no thinly dusted areas.

The copper piece is now ready for firing and it can be fired in the kiln or placed on the grid and fired with the torch. A spatula or lifting fork is used to lift the copper from the paper to the kiln or to the grid. The paper is lightly creased between the thumb and fingers to form a slight depression under the copper. The

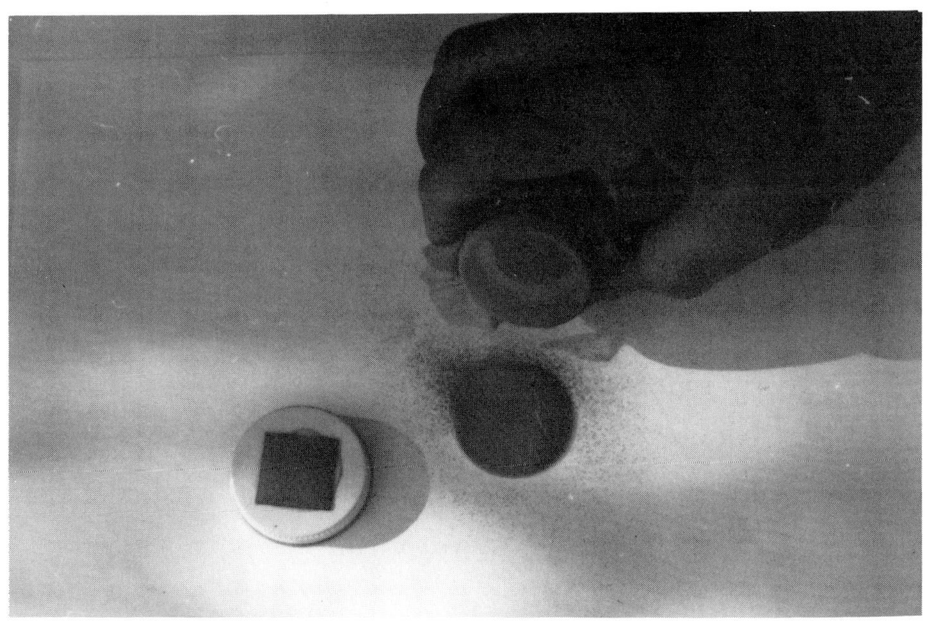

Dusting the enamel onto a copper piece

depression provides the space necessary to allow for insertion of the spatula and removal of the piece.

Firing takes but a few minutes in a preheated kiln. The piece should be watched carefully and removed when the enamel becomes fused and smooth. If the piece is fired with the torch, the enamel is fused by heating it from the underside of the grid.

The work is now allowed to cool. It is then put facedown into the pickling solution, where it is allowed to remain for three or four minutes. The pickling solution dissolves any firescale that may have accumulated on the back of the piece. Excessive pickling should be avoided, because it dulls the enameled surface.

The underside of the work should be kept free from particles of enamel, as they could be transferred to the kiln. Particles of enamel that may have become fused to the back of the piece will not be removed by the pickling solution but can be removed by using an abrasive cloth or a file.

Enameled piece immersed in the pickling solution

Dusting Aids

The circular aid is made by placing a fly screen over one end of a large fruit jar lid. A hose clamp is used to secure the screen. Be certain that the raw edges of the screen are not exposed.

The purpose of this device is to prevent the enamel from becoming attached to the bottom of the work. Place a clean piece of paper beneath with the work on top of the screen. The excess enamel will collect on the paper and may be returned to the jar with ease.

The other instrument is used in the same manner, except that there is no screen. The work is supported by the long legs; this allows the excess enamel to fall down to the paper beneath. This requires an 8-inch piece of ⅛-inch copper tubing that has been flattened.

Dusting Aids

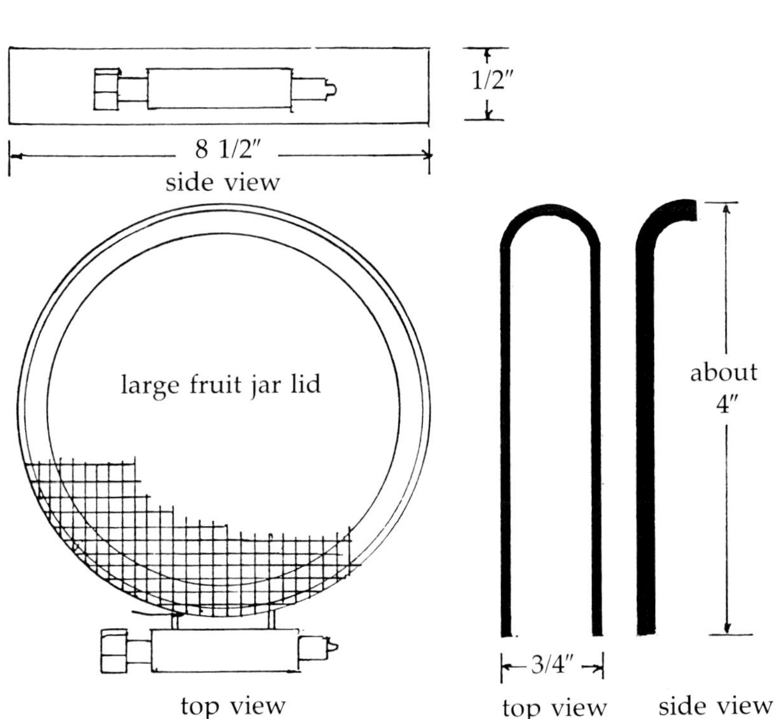

CHAPTER FOUR

Counterenameling

Copper expands upon heating and contracts upon cooling to a greater degree than its enameled surface, creating tension that can cause warping, cracking, or crazing of the article. In order to strengthen the piece and avoid these undesirable features, the piece can be counterenameled on the underside. Counterenamel is dry powder enamel; it is available from most enamel suppliers. It is usually gray in color and appears to be a mixture of many colors. Any of the regular enamels can be used as counterenamel, and often the enamelist combines his leftover enamels and uses them for counterenameling.

Counterenamel is usually applied to the back of the piece and then fired before any enameling is done on the front. It is necessary to prevent firescale from forming on the front of the piece during the firing process. A semiliquid product called Scalex is used for this purpose. It is applied with a brush and allowed to dry thoroughly before the counterenameling is done.

Counterenameling Procedures

1. Both sides of the copper are thoroughly cleaned with "00" steel wool.
2. Scalex is applied to the face side and allowed to dry.
3. The back is brushed with 7001 enameling solution, and a uniform coating of counterenamel is applied by the sifting method.
4. The piece is placed in the kiln with the counterenameled side uppermost and fired at a temperature of about 1400°F.

Placing counterenameled piece in the kiln

Counterenameled piece on the rack

5. The piece is removed from the kiln and placed in a safe place to cool.
6. After the Scalex has flaked off, the face side is cleaned with steel wool.
7. The face is brushed with 7001 solution and dusted with the desired color of enamel.
8. The piece is centered on a trivet with the counterenameled side down. The trivet is used for the purpose of preventing the counterenamel from sticking to the hearth of the kiln.
9. The copper piece with the supporting trivet is placed in the kiln with lifting fork and heated for a few minutes at a temperature of about 1500°F.
10. When properly fused, the article is removed from the kiln and cooled.

It is not always necessary to counterenamel. In fact, in some instances counterenameling may be inadvisable. This is especially true when small pieces of jewelry require that findings be soldered to the backs of them. Solder will not adhere to any enameled or counterenameled surface and can only be successfully used on freshly cleaned metal.

CHAPTER FIVE

Swirling or Scrolling

Swirling, sometimes referred to as scrolling, is a technique whereby several different colors of enamel are swirled together to create a random design. Swirling is one of the simple ways of obtaining interesting and satisfactory results with little previous experience. It provides a certain amount of enameling exercise, and the finished product is pleasing and intriguing.

Equipment Needed

1. The acetylene torch is used for swirling, as it provides excellent control of the temperature and facilitates the manipulation of the colors.
2. A grid is necessary. It provides a place upon which to set the copper piece when it is being fired with the torch.
3. A swirling rod is needed. The steel swirling rod that is usually furnished with the average enameling kit is not completely satisfactory. It has a tendency to stick to the molten enamel and requires frequent quenching. A better swirling rod can be made by the craftsman by sharpening a two-inch length of twelve-gauge copper wire and hard soldering it to an eight-inch length of stainless steel welding rod. Copper is an excellent conductor of heat and therefore a rod with a copper tip requires less quenching attention.

Swirling Procedure

1. The face side is thoroughly cleaned with "00" steel wool.
2. A thin coat of 7001 enameling solution or other binder is applied to the surface. The binder should completely cover the piece; otherwise, the enamel will pull away from the edges on firing.
3. A layer of enamel is dusted over the piece. It can be fired at this time, but it is not necessary to do so. The choice of colors is rather important and should be given careful consideration at this stage of learning.
4. Small lumps and threads are scattered over the enamel.
5. The spatula is used to place the piece on the grid.
6. The torch is lighted and heat is applied to the piece from the underside of the grid. The temperature is slowly increased to a point where the enamel is melted and the surface is smooth.
7. A random design is created by dipping the rod into the molten enamel and swirling it.
8. The work is slowly heated to a glowing red. This additional firing levels out the enamel and prevents cracking.
9. The piece is cooled and placed in the pickling solution to remove the firescale. The edges are filed and the back is polished and lacquered.

The enamels are darker when they are in the molten stage. Some colors lose their identity, and it is difficult to distinguish one color from another. The light and dark colors are more distinguishable and should be used together. Swirling is a random technique and cannot be entirely controlled. Overswirling should be avoided; the final results are more beautiful when the design is simple.

Placing threads and lumps on a copper circle that has been dusted with enamel

Starting the swirling act

The finished swirl

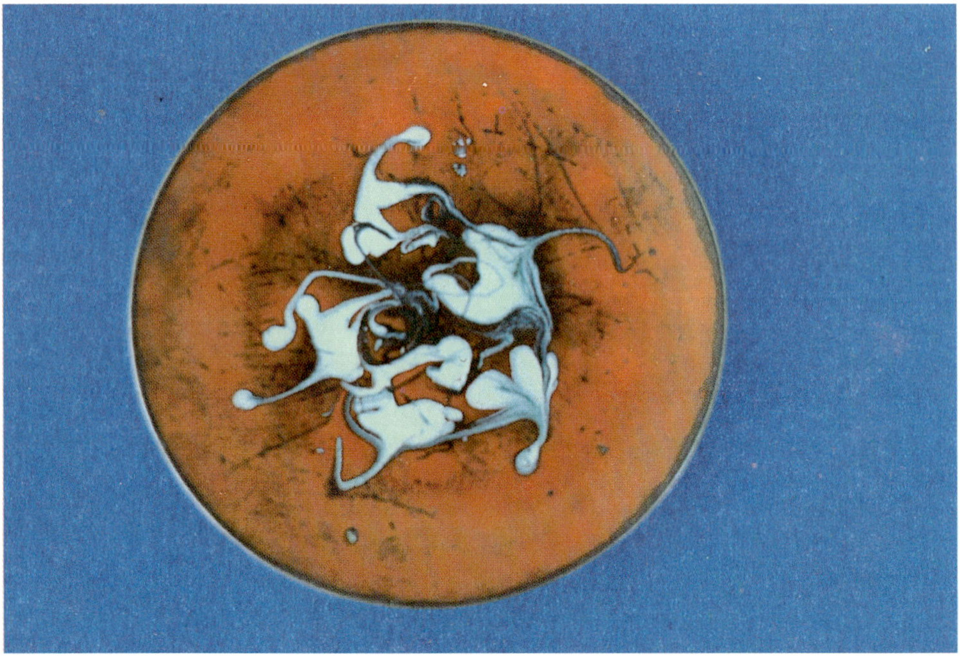

Finished piece of swirling

Example of swirling piece

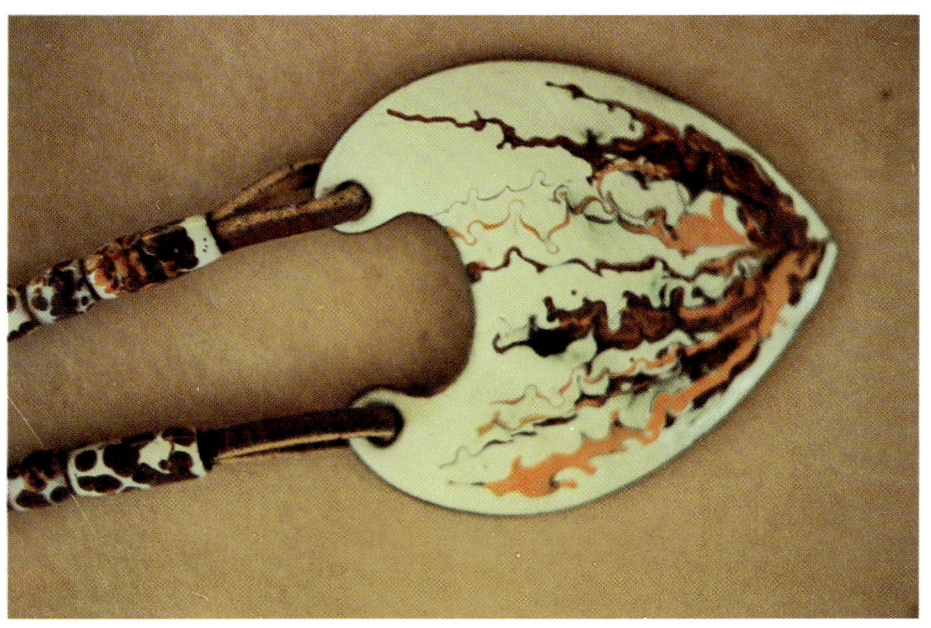

Another example of swirling piece

CHAPTER SIX

Crackle Enameling

Crackle is composed of finely ground enamel powder, a certain clay product, and water. The main purpose for using crackle is to obtain special, interesting effects over prefired enamel.

Crackle is purchased in semiliquid form and is obtained in one-ounce jars or in larger jars. The choice of colors is somewhat limited, but there are sufficient colors to satisfy most tastes.

Crackle may be applied by brushing, dipping, or spraying. When crackle is fired over regular enamel, the crackle shrinks and pulls apart in places, causing cracks or checks to appear. The undercoat is exposed through these cracks, resulting in a contrasting crackled effect. Except under certain circumstances (which will be explained later in this chapter), these cracks are random.

The piece being crackled need not be counterenameled, but counterenameling will help to strengthen it and provide assurance that it will be less likely to crack at some future time. Cracking or chipping of the enamel most often occurs when the work is enameled on one side only.

The student should confine work with crackle to minor pieces until he has perfected the technique. Satisfactory results are not obtained when small, flat pieces of copper are used. It is advisable to use domed or convex pieces that are about one to one-half inch in diameter.

Procedure for Simple Crackle

1. A fairly heavy and even coat of crackle is applied over a prefired coat of regular enamel.

2. The piece is placed aside to dry under a heat lamp or in some other warm place. It must be allowed to dry thoroughly; otherwise, when it is fired the moisture in the crackle will produce steam, which will, in turn, cause blisters.
3. The piece is then fired in the kiln at a temperature of about 1500°F, the temperature being maintained until the desired amount of crackle is produced.
4. The piece is removed from the kiln and allowed to cool. If the piece is counterenameled, it needs no further work. If it has not been counterenameled, it is placed in the pickling solution to remove the firescale and then it is cleaned, polished, and lacquered.

Procedure for Controlled Crackle

The procedure is the same as for simple crackle except that between the second and third steps an extra step is added. A design is scratched into the dry crackle, thereby exposing the undercoat. To a degree the random crackled effect is altered because the crackle has a tendency to follow the lines of the design.

In some cases, removal of large portions of the crackle may be desirable to effect a bold or definite pattern. This can be accomplished with any sharp instrument, such as a palette knife.

The firing and finishing is done as explained under the procedure for simple crackle.

Applying the crackle enamel to a fired undercoat

Finished piece of regular crackling

Sample of controlled crackle

Another example of controlled crackle

Example of controlled crackle

Controlled crackle sample

Example of controlled crackle

Sophisticated crackle

CHAPTER SEVEN

Findings, Soldering, and Cementing

Findings

Included under the heading of findings are items such as pin backs, ear wires, jump rings, chains, and other miscellaneous acccessories that are used in the construction of jewelry. These articles are usually manufactured and are not handcrafted. Much tedious work can be avoided by purchasing them ready-made.

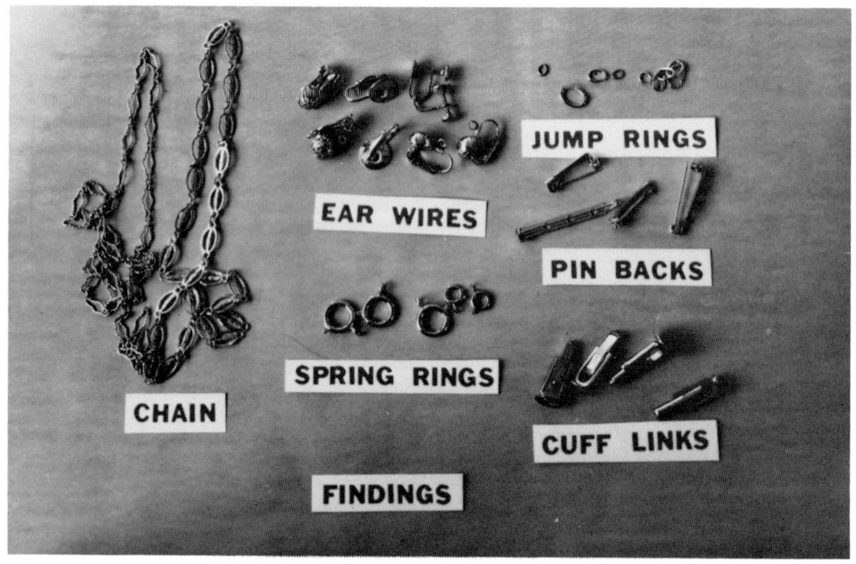

An assortment of findings

Findings are obtainable in sterling silver, fourteen-karat gold, and copper. They are also supplied in several plated finishes. Pin backs, for instance, can be had in silver-plate as well as in gold, rhodium, and nickel-plate. Some findings, such as cuff link backs and ear wires, are purposely provided with large pads in order to be readily cemented. Others are manufactured with small pads and are specially designed to be soldered.

Soldering

The technique used in soldering is quite simple, requiring two essential materials: soldering flux and solder. Soldering flux is a paste that can be obtained in small cans. Solder is composed of half tin and half lead and is purchased in the form of round wire. Since round wire tends to roll, it is wise to flatten it. This can be done with a hammer or by rolling it between jeweler's rollers. The thin ribbons of solder can be cut to various sizes and shapes with scissors.

Before a piece of jewelry is soldered, the firescale should be removed by immersing the piece in the pickling bath.

Hard soldering is sometimes listed as silver blazing, silver soldering, or just plain brazing. Hard soldering should not be confused with welding. Welding involves the fusing of parent metals. This is called *co*hesion, while any form of soldering is *ad*hesion.

The underside is then polished with steel wool until clean and free of any foreign material. A brush is used to apply a small amount of soldering flux to the space that is to be covered by the finding and also to the underside of the finding. Next, in sandwich fashion, the parts are assembled into one unit. The enameled piece, face down, is the bottom layer. A small ribbon of solder is the middle layer, and the finding is the top layer. The combined unit is then gently heated with a torch or placed on top of a hot kiln cover for a few seconds. Either method requires that the heat be applied to the enameled surface. If the torch is used, the article is placed enameled side down on the metal grid and heated from the underside. Extreme care must be exercised, since overheating could damage the finish or cause the enamel to stick to the grid.

Heat will eventually melt the solder but will not damage the enamel if the torch is held at a distance and the flame turned low. Before long the flux will begin to boil and the solder to melt and flow between the pieces. At this point the article is removed and allowed to cool. The solder soon becomes rigid and the finding is securely held in place.

Soldering is a term used to describe a process where two or more parent metals are joined with the use of a lower-melting alloy. The word solder is a derivation of the Latin word *solidare,* which means "to make solid."

The soldering process is classified under two headings: soft soldering and hard soldering. Any soldering that is accomplished below 850°F is labeled soft soldering. That which is done above 850°F is called hard soldering.

Soft solders are usually alloys of tin and lead. They are grouped by weight, and more than one term may be used for this same alloy: "50/50" and "half and half." There are many other alloys of tin and lead that are used for soldering, such as 40/60, 30/70, and 60/40. They all melt within a few degrees of each other, but each has a special usage. The melting point of 50/50 is about 358°F.

Fluxes

The use of fluxes is absolutely necessary, whether for soft or for hard soldering. When a piece of metal is heated, it tends to oxidize, forming a surface that is difficult to solder. Fluxes are deoxidizers and act as agents to dissolve the oxides as they form. Each type of soldering requires its own special flux. If 50/50 solder is to be used, the best flux is a paste form. This is especially true where small findings are soldered to copper.

When silver brazing, the best flux to use is the one recommended by the manufacturer of the alloy being used. Silver-brazing alloys are manufactured in many types, and are classified according to the amount of silver used. A very excellent flux, made by the Harris Company, called Stay Silv #5 is recommended for the enamelist.

There are four essential factors to be considered for any kind of soldering. They are clean work, good solder, proper temperature, and proper flux.

Cementing

A cemented joint is not as strong as a soldered joint. Cementing, notwithstanding, is a very reliable method for attaching findings and is employed much of the time. If the piece has been counterenameled it must be cemented because solder will not adhere to enameled surfaces.

If the back of the article has not been counterenameled it should be cleaned with steel wool before the finding is cemented to it. A small amount of epoxy cement is placed on the underside

Example of soft soldering on Brittania metal. Each grape was soldered in place.

of the finding. The finding is then positioned, pressed down firmly, and set aside for several hours to harden.

Cleaning will be necessary if soldering has been done and probably will not be if the finding has been cemented. As a finishing touch, the entire back can be lacquered.

CHAPTER EIGHT

Controlled Design

This aspect of enameling offers an exceptional opportunity for self-expression and using one's imagination. The beginner will wish to explore the many possibilities, but instructions will be given here for one specific example of controlled design. The poinsettia flower employs the typical controlled design technique and is a good illustration of what can be done. The necessary steps used in making the poinsettia merit strict attention if the final results are to be successful. These steps are as follows:

1. A 1½-inch copper disc or circle is used for this purpose. First the copper is cleaned and a thin coating of 7001 enameling solution is applied. A modest coat of opaque white enamel is carefully dusted over the piece.
2. Ten to fifteen irregular-shaped, opaque red lumps about the size of this capital O are selected. They are placed in a circle midway between the center and the edge of the copper disc. Care must be exercised to avoid disturbing the undercoat, as it has not as yet been fired.
3. In the third step the work is placed on the grid and from the underside it is heated with the torch until the lumps are fused and soft enough to be moved about. A copper-tipped swirling rod is then used to pull the other edge of each individual lump toward the rim of the disc. The torch flame should be used continuously in order to maintain the necessary temperature. A little twist of the rod will assist in shaping the outer half of the petal. When a stainless-steel rod is used, it should be quenched frequently in cold water to prevent it from sticking to the enamel.

Preparation for controlled design poinsettia flowers

Using copper-tipped rod to pull lumps outward, forming petals

Pulling the petals inward

Completed poinsettia

Less quenching is needed when a copper-tipped rod is used, the construction of which is explained in Chapter Five.

4. The lumps are pulled into the center, making them appear more like poinsettia petals. This procedure produces an accumulation of unwanted enamel.
5. A stainless steel rod is heated red hot and dipped into the accumulated enamel, causing the enamel to adhere to the rod. The rod is lifted upward and a thread of enamel is drawn. The torch flame is used to separate the thread from the work.
6. The sixth step provides for the addition of the yellow center particles. A lump of yellow enamel is removed from the assortment, wrapped in a heavy piece of paper, and with a hammer is broken into small fragments. About a dozen of these tiny fragments are selected and placed close

Poinsettia treatment on a tray

Poinsettia treatment on tile

together in the center of the flower. The work is reheated until all the enamel is level and smooth.

7. At this stage the final cleaning and finishing takes place.

This method of enameling requires considerable dexterity and special skills. These can be developed in due time with practice. After the beginner or novice has acquired the necessary confidence in his ability to manipulate the tools, there are many new and exciting avenues to follow.

Tray with tile, using lumps

Controlled design technique

Controlled Design enameling offers many opportunities for expressing one's imagination. The results are usually pleasing and beautiful. The choice of colors is an important factor to be considered. The use of contrasting colors will add to the character and quality of the finished product.

The copper-tipped swirling rod is an indispensable tool for this type of enameling. The directions for making this implement are listed here with accompanying illustrations.

The required materials are as follows:

1. A two-inch length of #12 gauge copper rod and
2. A nine-inch length stainless steel rod 1/8th inch in diameter. Inexpensive fondue forks are ideal for this purpose. They have a wooden handle to protect the fingers from heat and are usually made of stainless steel. The fork at the end is sawed off, and the end of the rod filed smooth and even.

Controlled design with the use of lumps

Controlled design method. Tray with small tile.

Another example of controlled design

The copper rod is also prepared by sawing one end to make it straight, smooth, and even. There must be a good match at the joint of the two pieces. The two pieces are placed end to end on charcoal blocks so that the ends are matching. Small staples are used to secure the metal parts to the blocks. A piece of wire can be fashioned for this, or a part of a paper clip can be utilized. The next step is to add a small amount of flux to the end of the steel and fuse a small piece of hard solder at this point. Close the gap between the two parts and reheat until the solder flows into the joint. Put a little pressure on one of the charcoal blocks to cause the joint to close and adhere. Allow the work to cool. Sharpen the copper on the point.

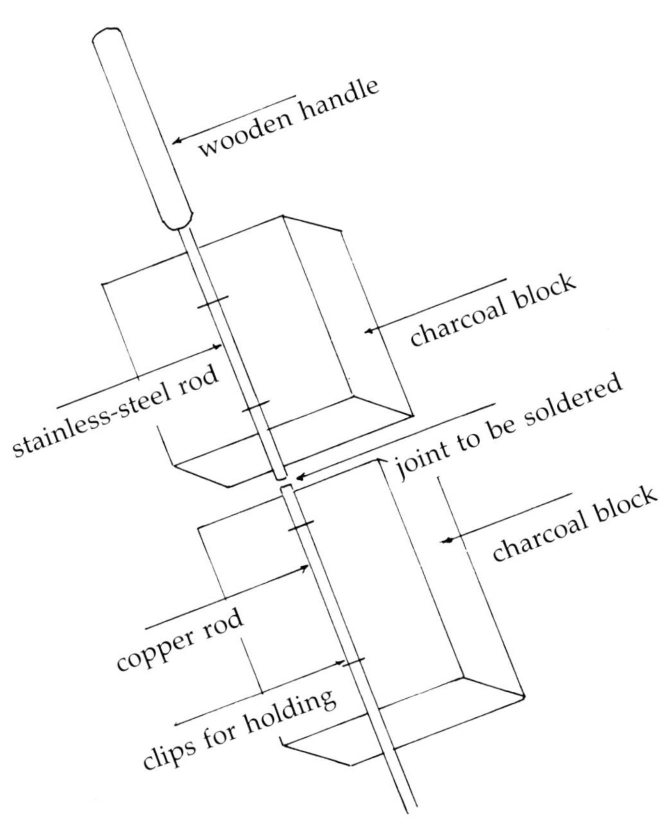

Controlled design on matching jewelry

Controlled design treatment

CHAPTER NINE

Stenciling and Masking

Stenciling and masking are closely related and are important phases of the enameling art. They offer distinct possibilities for creativeness. When a stencil is made, a design is cut out, leaving open spaces and lines through which the powdered enamel is sifted. Masking, on the other hand, is the reverse process: the design is formed by covering certain areas and dusting the enamel between and around them.

In most instances the stenciled or masked article is fired in the kiln rather than by means of the torch. When the torch is used the article cannot be counterenameled.

If the student wishes to counterenamel, he must do so before the stenciling or masking is started. If he chooses to forego counterenameling he can finish the back of the work by cleaning, polishing, and lacquering it.

The Open Stencil

A design is drawn or traced on a sheet of tracing paper. A scissors, X-acto knife, or razor blade is used to cut out the stencil. If a knife or razor blade is used, the work should be placed on a sheet of glass. This will make the cutting easier and there will be less tendency to tear the paper.

The face side of the work is enameled with a background coating. It is brushed with oil and the stencil is carefully placed in the proper position on the enameled piece. A selected color of enamel is sifted over the stencil, building up enough powder to satisfactorily cover the open spaces. The stencil is then carefully

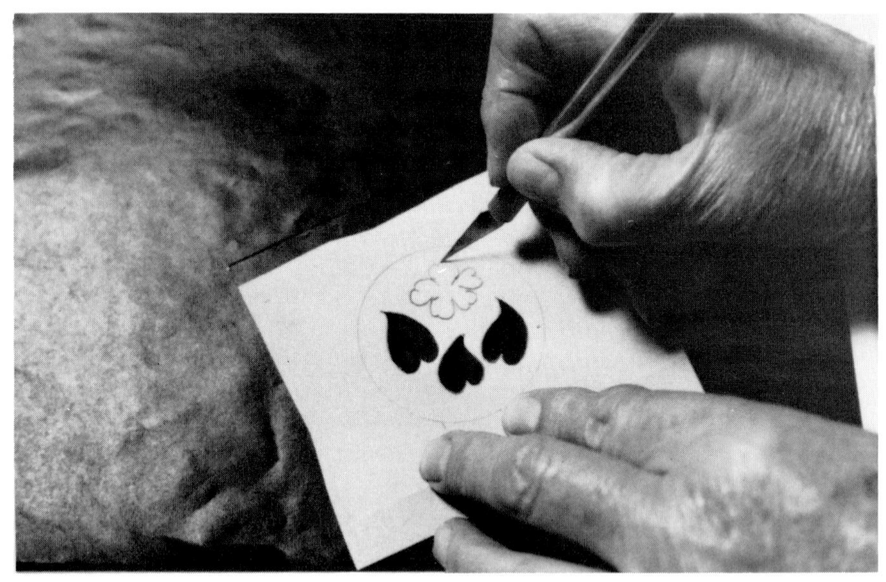

Cutting out stencil with knife

removed, exposing the design under it. At this point the work is ready to be fired.

Sometimes a stencil is needed for a curved or flared piece such as a tray or bowl. It is difficult to make a piece of paper conform to the curved surfaces of a bowl; therefore, slits are cut in the edges of the paper, making it possible to overlap the sections. This allows the stencil to take the shape of the bowl. It is pressed into the bowl and the overlaps are taped to secure them. Then it is removed and the stencil design is sketched in and cut out.

Simple Masking

When simple masking is done, the entire piece is enameled and then portions of it are covered with a masking material. In this way colors under the masked portion are retained when a second coat of enamel is sifted over the piece. The masking material must lie flat on the enameled surface. This is necessary in order to prevent the powdered enamel from falling under the edges.

Natural objects such as leaves, ferns, and moss can be successfully used for masking. Paper cutouts, either original or traced, are probably more often used because they are easier to handle and can be made to depict any desired motif. Flowers, birds, boats, trees, and animals are among the favorite cutouts, and excellent results can also be obtained by using geometric and freeform designs.

Directions for doing simple masking are as follows:

1. The copper is cleaned and enameled, first coat.
2. After a thin coat of 7001 solution has been applied to the entire surface of the work, the masking material is placed in position.
3. The mask is brushed with 7001 solution, careful attention being given to the edges.
4. The enamel is sifted over the exposed spaces. The student must make sure that they are uniformly covered. Any enamel that accidentally falls on the mask is not objectionable; it will cling to the oiled surface.
5. The mask is carefully removed and the article is ready for firing.

Sophisticated Masking

When the design includes several areas, each of a different color, a more elaborate procedure is followed. Each area is treated individually, with its appropriate color and separate mask. Finally, when all areas are covered with enamel, the firing can take place.

Let us discuss the scene depicting a light blue sky, snow-covered mountains, a blue lake, and details such as rocks, trees, and clouds. The foreground is in two or more shades of green. A special technique is necessary, and certain steps must be observed to assure results that are satisfactory. They are as follows:

1. The outside dimensions of the mask are determined by laying the selected piece of copper on a sheet of paper

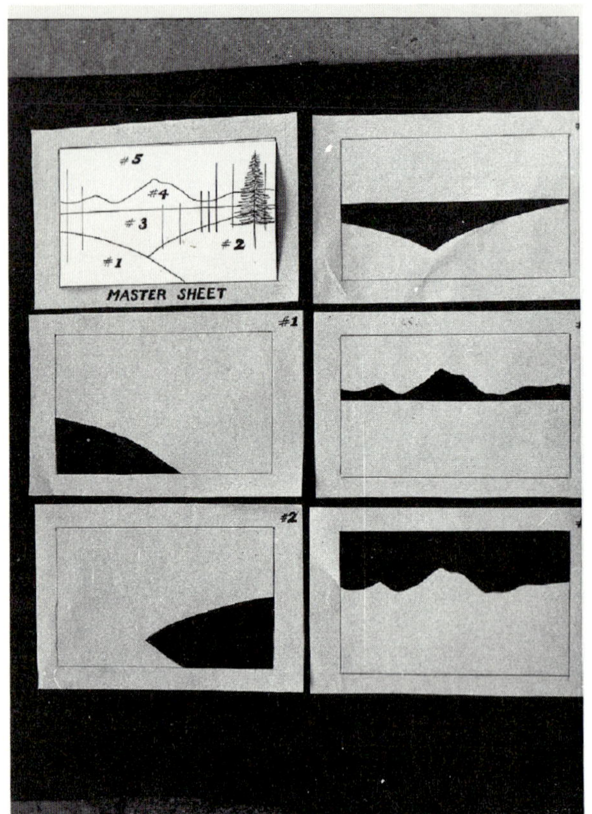

Making a sophisticated mask

Final results, using a sophisticated mask

and tracing around it. This sheet of paper will be referred to as the master sheet.

2. In the outlined space on the master sheet the scene is sketched. Be sure to include all the details such as the trees, rocks, and the snow on the mountains. Each area of the scene is labeled with a number that indicates the order in which it will receive treatment. The foreground will be the first to be covered with powdered enamel, so it should be #1. If there is more than one part to the foreground, each part should be numbered. The scene shown here has two parts to the foreground; therefore, the left side is #1 and the right side is #2. The lake is #3, the mountains, #4, and the sky, #5.
3. Five copies of the master sheet are made on tracing paper, omitting the numbers in the sections and omitting minor details.
4. These tracings are numbered from 1 to 5, and the numbers are placed in the upper–right-hand corners of the border.
5. Number 1 tracing should have area #1 or the left portion cut away, leaving in the mask all but this part. Number 2 tracing should have area #2 or the right foreground portion removed. Number 3 should have the lake section cut out and removed, number 4 the mountains, and number 5 the sky. Each tracing should have a section cut out corresponding to its number on the master sheet.
6. With the five separate masks satisfactorily made, we can now proceed to do the enameling. The copper base is cleaned and a coat of 7001 solution is brushed over it and also over the face side of the five masks.

Mask #1 is placed in position on the copper and held firmly in place with one hand while the selected green enamel is sifted over the exposed portion of the foreground. The mask is then carefully removed, leaving the enamel undisturbed. Next, mask #2 is gently placed over the copper, and in a similar fashion the exposed area in it is covered with the other green enamel. The mask is removed, and the same procedure is followed with each mask until the entire piece of copper has been covered. Any imperfections in the work can be readily eliminated by the application of a slight pressure with the palette knife.

The snow can be added by picking up a small amount of white with the palette knife and placing it on the mountaintops. Small patches of white can be dropped in the sky for clouds. By using threads of a contrasting color the tree trunks can be added.

The work is now ready for the first firing and should be placed in the kiln and fired at about 1500°F. When it is smooth and completely fused, it is removed and allowed to cool.

At this time the foliage is added by picking up a small amount of enamel on the edge of the palette knife and gently dusting it in place. Any other fine details are also taken care of. A bit of chartreuse or other light color can be added to highlight the trees, and darker shades can be put in for shadows.

With everything completely finished, the work is now ready for the second and final firing.

COLOR NUMBERS FOR ENAMELS USED IN THE SCENE*

Listed in order of their application

First Firing

Front foreground	No. 100 pea green
Second foreground	No. 577 jade green
Lake	No. 866 cerulean blue
Mountains	No. 595 dove
Snow on mountains	No. 118 white
Highlights on snow	No. 305 clover
Mountain shadows	No. 304 mauve
Sky	No. 316 mist blue
Clouds	No. 118 white
Highlights on clouds	No. 305 clover
Tree trunks (tapered threads)	No. 301 caramel

*These numbered enamels are borrowed from those listed in the *Thompson Catalog*.

Second Firing

Tree foliage	No. 688 spruce green
Highlights on trees	No. 736 chartreuse
Tree shadows	No. 137 rejane green
Flowers in foreground (30 mesh lumps)	No. 1068 citron yellow

Another example of masking

Example of stenciling enameling

Stenciling design with fine-line-black

Design effect made by shifting stencil

CHAPTER TEN

Restoration

The torch is used in the restoration of work that has been rejected or discarded because of chipped or cracked enamel, displeasing colors, overfiring, or other imperfections. In many instances the copper piece is worth salvaging. If the piece has not been counterenameled and is small enough to be heated with a torch, something can usually be done to restore it.

Let us first consider the piece that has been sparsely enameled and has too thin a coating. It may be restored by simply applying additional enamel. A sufficiently heavy coat of any opaque color will completely obscure the undercoat. The addition of a transparent color will let the undercoat show through, sometimes producing a desirable color effect.

When the enamel is too thick, it is possible to remove some or all of the unwanted portion. To do this, a stainless-steel rod with a sharpened point is used. The work is refired and the point of the rod is heated red hot and dipped into the fused enamel. When the rod is firmly attached to the enamel, a small amount of it is removed by lifting the rod and pulling out a thread. The thread is then drawn over to the edge of the piece and is melted off with the flame. More enamel can be removed in the same way, the process being repeated as many times as is necessary.

If the background color is acceptable but the composition or design is displeasing, the problem may be solved by adding some threads or lumps and swirling them. In this way a fresh and beautiful effect can be achieved.

An overfired piece can usually be restored to its proper color by refiring with an additional coat of the same enamel. Often low-firing enamels that have been overfired can be restored by simply refiring at a lower temperature.

Technique to remove enamel

When a discarded piece is unacceptable because of cracks or chipped edges, it may be repaired. A torch is used to heat the piece to a bright red, and a copper spatula is used to carefully spread the melted enamel over the damaged places. The piece is then refired in order to level the enamel.

Occasionally it may be desirable to completely remove the enamel. This may be accomplished by using either the kiln or the torch. The rejected article is heated until it becomes red hot and then is immediately plunged face-downward into ice water. This will cause most of the enamel to crack off. It may be necessary to repeat this treatment in order to remove all of the enamel.

There is still another method that may be used in removing unwanted enamel. The unsatisfactory piece is placed on the grid and is heated to a bright red color. This high temperature must be maintained while the enamel is being removed. A stainless-steel rod is laid flat on the surface of the piece and is pushed across the piece, gathering up the enamel. When the accumulated enamel reaches the edge, the rod with the adhering enamel is lifted off and plunged into cold water. The quenching action causes the enamel to crack off, leaving the rod clean.

A thin film of enamel may remain on the work. This is not usually objectionable, as it can serve as the initial coat when the piece is redecorated.

CHAPTER ELEVEN

Foil Overlay

Spectacular effects are obtained when foil overlays are embedded in enameled pieces. The following statements relate to foil overlays.

1. Gold, silver, and copper foils are most commonly used. Copper is used more extensively than either of the other two because it has the advantages of being thicker, less expensive, and easier to handle.
2. Copper foil is obtainable in hobby shops and is sometimes called tooling copper. Thirty-six gauge is ordinarily preferred, but lighter metal is also suitable.
3. Gold and silver foils are seldom found in the ordinary hobby shops. Usually they must be ordered through a catalog. They are manufactured in sheets, each approximately four by four inches and separated by protective papers.
4. Either opaque or transparent enamel may be applied to copper foil overlay. It is recommended that transparent be used on gold or silver, as this permits the beautiful luster of the metal to be seen through the enamel.
5. Tweezers are needed when working with foils. This is especially true in the case of gold and silver foils, since they are very fragile.
6. A small, sharp scissors is needed to cut the foil overlays.
7. A torch or kiln can be used to fire the piece that is overlaid with foil.
8. The edges of the overlay may curl up when the piece is being fired. This can be controlled with a steel rod or spatula.

Copper Foil Overlays

Copper foil must be cleaned on both sides with steel wool, as both sides make contact with enamel.

In most cases the design for the overlay is first sketched on a sheet of tracing paper. The paper is then glued to a piece of foil with white water-soluble glue. After the glue has dried sufficiently, the design is cut out and placed in water to dissolve the glue and permit the removal of the paper.

If one desires, the design can be cut out freehand from the foil, or it can be sketched directly on the foil and then cut out.

There are two methods used in applying foil overlay.

First method:

1. The foil only is oiled with 7001 solution and covered with powdered enamel.
2. The foil is placed on the previously enameled surface.
3. The entire piece, including the foil, is fired.

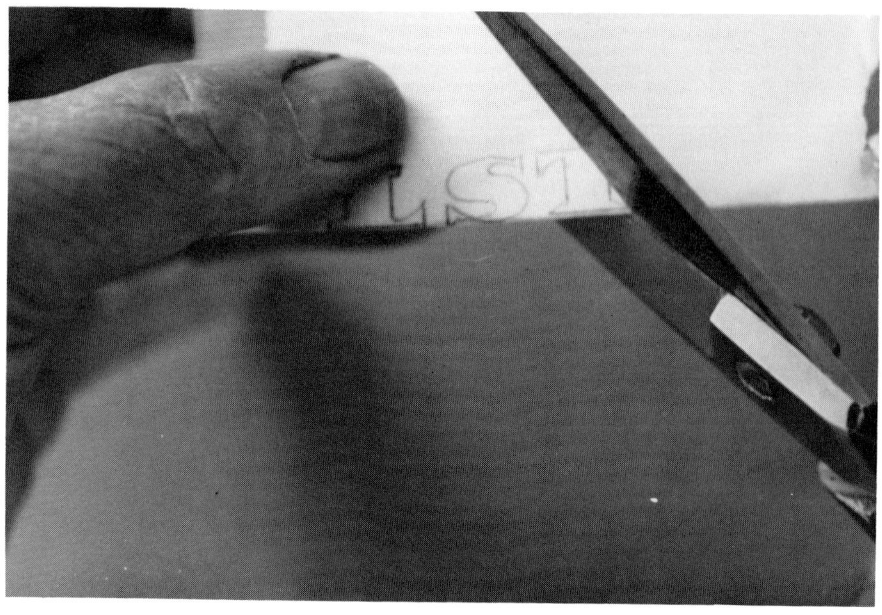

Foil wrapped in tracing paper

Butterflies of silver foil, oversprinkled with transparent enamels of blue, yellow, and green

Second method:

1. The foil is oiled and placed on the previously enameled surface.
2. Enamel powder is then sifted over the foil.
3. The entire piece, including the foil, is fired.
4. When opaque enamel is used, special care should be taken to remove any scattered particles that may fall on the undercoat.

Where transparent enamel is used as a background, beautiful effects can be achieved by purposely sifting the enamel over the background as well as over the foil. Clear flux can be used in the same way, and the natural color of the polished copper will be revealed through it.

Gold and Silver Foil Overlays

Since gold and silver foils are thin and delicate, much care must be exercised to prevent them from folding and wrinkling. Any distortion may show in the finished piece, and this should be avoided if possible. The foil is most efficiently handled when it is sandwiched between the two sides of a folded piece of tracing paper. The design is sketched on the outside of the folded paper and the foil is slipped down to the folded edge, where it is held securely. This technique protects the foil and keeps it in place while the design is being cut out. It also eliminates the risk of fingerprints.

Tweezers are used to carefully separate the foil from the paper. The overlay is then pricked with a needle in several places. This procedure forms vents for the purpose of allowing trapped air to escape when the piece is fired. Now the foil overlay is picked

Butterfly of silver foil decorated with fine-line-black

Tree trunk of copper foil. Blossoms are white lumps.

Birds and flowers of copper foil. Blossoms are floral wafers.

Bird of copper foil, with stenciling

Leaf of ceramic gold

up with the tweezers and placed in position on the oiled surface of the enameled background. The overlay is also oiled and smoothened.

If flux is used, an interesting effect is achieved by revealing the natural color of the foil. When transparent colors are used the gold and silver show through, but the color will be primarily that of the selected enamel.

Special attention is required when gold and silver overlays are fired because of their comparatively low melting points.

CHAPTER TWELVE

Cloisonné

Cloisonné is one of the older forms of enameling. It was done by the Greeks more than two thousand years ago. Today, however, the techniques and methods are different. Enamels are greatly improved and the enameling process has been simplified.

The first step in doing a cloisonné is to select a piece of copper that is the desired size and shape. It is enameled on one side with a background coat of either transparent or opaque enamel.

One or more pieces of copper wire are cut and shaped to form a planned design. Twenty-gauge copper wire is used and is shaped with the aid of small round-nose pliers. The wire is hammered lightly on a flat steel surface. Hammering hardens and flattens the wire and produces a greater area of contact with the base.

The sections of wire are placed in the correct positions on the enameled piece, each section being held in place by a small amount of white glue applied at strategic points, such as the ends of all pieces and the centers of the longer ones. The glue is used sparingly and is applied with a small sharpened stick of bamboo or a toothpick.

The work is allowed to dry for several hours or until firm, and then it is fired to burn out the glue and to partially embed the wire in the base coat of enamel. Wait until the piece is cool before taking the next step.

The small enclosures referred to as cloisonnés are now filled with enamel. Each color is prepared as needed by placing a small amount on a sheet of glass and with a palette knife carefully blending in a few drops of Klyr-Fire. The consistency of this mixture must be just right, not too thick and not too thin, in order

16-gauge copper cross with decorations of cloisonné ribbon

Examples of cloisonné work

Cloisonné pendant with beads

Fish in bowl

Example of cloisonné work

Another example of cloisonné work

to properly and neatly fill the cloisonnés. After each section has been filled and packed with the proper color of enamel, it is leveled to the top of the wire.

Firing is now necessary. A temperature of 1450°F to 1500°F is maintained for a period long enough to completely fuse the enamel.

The enamel tends to shrink slightly each time it is fired. After the work has cooled sufficiently, a determination as to whether or not more filling and another firing is required can be made. Sometimes a cloisonné looks beautiful and complete, even though the wire is slightly higher than the enclosed enamel. If a level appearance is desired, more enamel and another firing will be needed.

The completed piece is pickled for a few minutes, after which it may be lightly polished to bring out the luster of the metal wire.

The enterprising enamelist who wishes to extend his efforts will find many possible variations in the art of cloisonné.

Silver may be used as a base metal instead of copper. When transparent enamels are used over silver a more brilliant effect will result.

Silver, gold-filled wire, or flat fine-silver ribbon may be used instead of copper wire.

Flat fine-silver ribbon is known as cloisonné wire and is often preferred to all other types of wire. It is normally eighteen-gauge wide and thirty-two–gauge thick. When it is placed upright on its narrow edge, deep recesses for the enamel are provided and fine lines and detail are achieved.

CHAPTER THIRTEEN

Thread Drawing

There is often a need for threads that are not commercially available. With the aid of a few tools and some special pieces of equipment, the enamelist is now able to make threads of all sizes and colors.

Thread making requires the use of an acetylene torch, which should be equipped with a #3 tip, the one most frequently used by the beginner.

A drawing rod is also necessary and can be constructed from a ten-inch length of 1/8-inch stainless-steel welding rod. A handle is made by doubling back the rod about an inch on one end. The other end is sharpened to a point by filing, grinding, or hammering.

A cup in which to fuse the enamel is needed. This should be made by the craftsman and should measure about 1¼ inches in diameter. A piece of pipe is necessary in the construction of the cup. The inside diameter of the pipe should be about 1¼ inches and the length about one inch. A two-inch square of eighteen-gauge copper is used to make the cup. The square of copper is placed on top of one end of the pipe. With a ball-peen hammer the central portion of the square is driven into the pipe, bending the copper square into the shape of a cup. A block of wood provided with a hollowed-out, cuplike depression can be used in place of the pipe.

Thread-Drawing Procedure

Approximately half a teaspoon of powdered enamel is put into the cup and the cup is placed on the grid. A slight depression

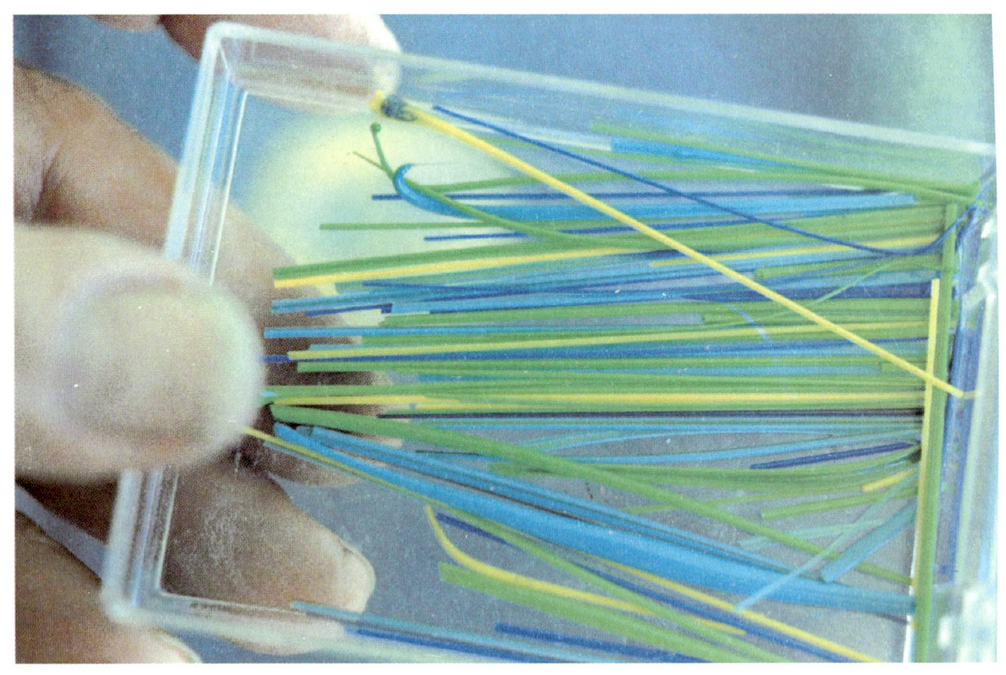

An assortment of threads of several sizes and colors

or a hole in the grid provides a convenient place to set the cup. The heat from the torch is applied to the cup from the underside of the grid. When the enamel has become uniformly fused, the point of the drawing rod is heated red hot and dipped into the center of the melted enamel. The rod is rotated a full turn in order to gather up enough enamel to hold it securely and prevent it from becoming disengaged. The rod is then raised, drawing with it a thread of molten enamel. If the cup tends to be lifted up with the thread, it is because the enamel is not as hot as it should be. When the thread has been drawn to the desired length, it is separated from the remaining enamel by carefully guiding it to the rim of the cup and melting it free with the flame. The rod and the attached thread are then placed on an insulated board to cool.

The thickness of the thread is influenced by the temperature of the enamel, the speed at which the thread is drawn, the type of enamel (hard or soft), and the size and shape of the drawing rod. A fine thread, for instance, will be obtained when soft enamel is used at high temperature and the thread is drawn rapidly. A coarse thread is produced when the enamel is barely melted and the thread is drawn slowly.

The cup must be cleaned before a new color is used. A satisfactory means of accomplishing this is to heat the cup red hot and then plunge it into ice water. Most of the enamel will crack off and any that remains can be chipped off with a hammer. The tip of the rod can be cleaned in the same manner.

Special Threads

Curved Threads

Curved threads can often be used to an advantage. They can be made by coiling molten enamel around the drawing rod. The rod is firmly embedded in the molten enamel and the thread is drawn. The rod is rotated or twirled in one direction, winding the molten enamel around it and producing a continuous coil. The curvature of the coil increases in proportion to the increase in the distance between the cup and the rod. If small curves are desired, the rod should be twirled near the cup. If large curves are wanted, the twirling should be done at a greater distance from the cup. After the coil has cooled it may be cut or broken into a variety of sizes. The curves will be irregular, but most of them can be used.

Bent Threads

Sometimes a bend in the thread is required. This is made by softening a small spot of the thread by holding it over a lighted birthday candle. The finer the thread, the less heat will be needed to soften it. A right angle will be produced when the thread is held horizontally above the flame. When the thread becomes soft enough, it will sag to a vertical position.

Beading

Beading is a term used to describe the balling-up or bunching effect that occurs when heat is applied directly to the end of the thread. The end of the thread is held above the flame of a match or small candle. The enamel fuses and crawls back on itself, producing an accumulation of material. This beading technique is a valuable asset in certain types of ornamentations, such as flower stamens and butterfly antennas.

Elongated Threads

A technique can be employed by which cattails, tear drops, barley heads, petals, and other similar ornamentations can be made. A thread is held horizontally over a candle flame. Tweezers are needed at this point to hold one end of the thread. The heat of the flame is concentrated on one spot of the thread. When the

Barley heads, made by the process of elongation

Cattails. Elongated by holding over candle.

Curved threads and tadpoles. Made by elongation process.

spot becomes soft and pliable, the thread is removed from the flame. A slight pulling tension is quickly applied to it from either end, causing it to stretch and elongate into a fine thread. The various sections can be separated and used for special effects such as stems, petals, cattails, teardrops, and barley heads.

Straight Threads

Straight threads are cut to a suitable size and used for making stems, buds, petals, and numerous other ornamentations. The butterfly is an example of what can be accomplished with straight threads. The following are specific directions for making a butterfly.

1. A 1½-inch copper circle is enameled with a background coat of transparent enamel over clear flux.

Straight threads, laid along outside edges and drawn in with copper rod

Use of curved threads placed carefully and drawn into center

Using straight thread as indicated in text

2. Four or six straight one-inch threads about 1/32 inch in diameter are needed for the butterfly wings. For best effect, more than one color of threads should be used. There should be two threads of each color, one for each side of the butterfly.
3. These pieces are arranged on the enameled circle as illustrated and held in place with Klyr-Fire.
4. The copper piece is then placed on the grid and heated with the torch. When the threads have become completely fused, they are drawn with a copper-tipped swirling rod to form butterfly wings, as shown in the second illustration.

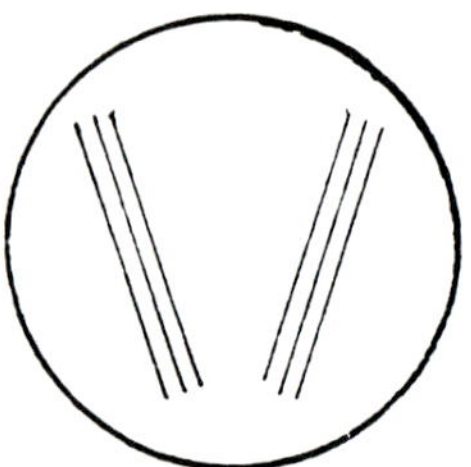

5. After all four ends of each group of threads are drawn toward the center, the middle section of each group is also pulled to the center. This causes an excess amount of enamel at the junction, and this excess must be removed. This is accomplished by heating a stainless-steel rod and dipping it into the center of the accumulated enamel and drawing a thread. The thread can be disconnected by applying heat to the base of it.
6. The antennas are made by using two short lengths of fine threads that have been drawn and beaded according to previous directions. A half-inch section of slightly larger thread is used for the body. These threads are held in their proper position with Klyr-Fire, or they may be placed on the work while it is hot, and then fused.

7. The work is refired and the final shaping of the wings is completed with a copper-tipped swirling rod.

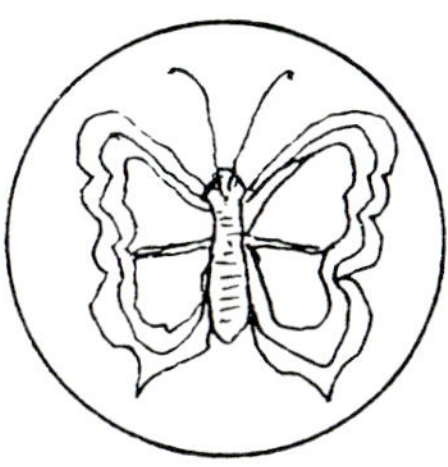

The finished butterfly will be similar to the one in the following illustration.

Using straight threads. Drawn in with copper-tipped rod.

CHAPTER FOURTEEN

Sgraffito

Sgraffito is a technique whereby a second coat of enamel is applied and then partially scratched off, exposing the undercoat and producing a particular pattern or design. Either opaque or transparent enamels are suitable, and they may be used separately or in combination. There are two popular methods by which this technique may be utilized with good results. The procedure is simple, but it is essential that the proper sequence be followed.

The first method should be used when a phantom effect is desired.

1. The piece of metal is thoroughly cleaned, and 7001 solution is applied. The work is enameled and fired in the usual manner.
2. When the piece is sufficiently cooled, it is sprinkled with a light coating of enamel (the coat is usually contrasting in color). No oil or binder is used, and the piece is not fired.
3. A stylus is used to sketch or scratch a design in the enamel, exposing the undercoat. If the enamel is too thick, it tends to accumulate ahead of the stylus, resulting in an irregular edge, which may be objectionable. A small sharpened stick of bamboo makes an excellent stylus and is preferable to a metal one.

   A piece of cardboard or heavy paper, cut to resemble saw teeth, can be zigzagged across the work, creating an interesting pattern. A comb can also serve effectively in the same manner.
4. The work is now ready for firing and finishing.

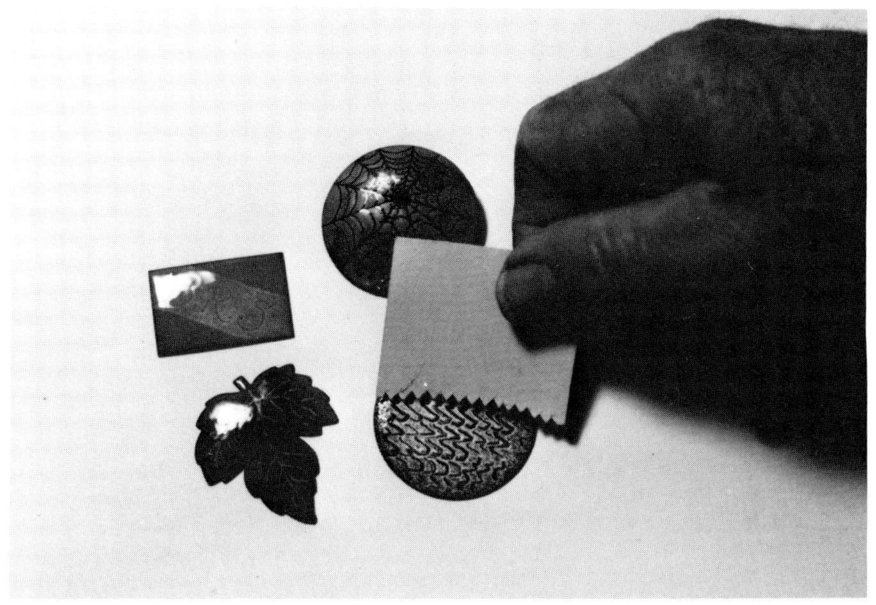

Designing with a paper comb

The second method is used when a more distinct design is needed or when larger areas of the base coat are to be exposed.

1. The copper piece is cleaned, enameled, and fired as in step 1 of the first method.
2. Klyr-Fire or gum tragacanth is sparingly sprayed or brushed over the preenameled surface and another coat of enamel is applied.
3. The piece is allowed to dry until a light crust is formed and the enamel clings to the binder. At normal room temperature this will require from one to two hours, but the time can be shortened to a few minutes by placing the work under a heat lamp.
4. When the enamel is sufficiently set, the design is scratched into the crust with a stylus. The unwanted portion of

This spider web was made using the sgraffito method

enamel is removed, and the undercoat is exposed in the appropriate places. Extreme care should be exercised to remove all loose particles of enamel.

5. When satisfactory results have been achieved, the work is ready for firing.

If firing in the kiln, the article can be counterenameled before any other work is done. If the firing is done with a torch, the counterenameling must be omitted and the back cleaned, polished, and lacquered.

Sgraffito is a technique that can be used advantageously for making initials, monograms, or script writing.

CHAPTER FIFTEEN

Enameling on Silver-Plated Steel

Silver-plated steel provides a good surface upon which to enamel, and the results are usually rewarding. There is one factor to consider: the choice of precut shapes and sizes is limited. Those that are available are nicely designed and cost about the same as copper pre-cut pieces.

Silver-plated steel is plated on both sides, the front side having a spangled pattern that adds to the beauty of the finished piece. Transparent enamels should be used on silver-plated steel, as opaque enamels defeat the purpose of the silver under the enamel. Any opaque enamel will hide the silver and destroy the beauty of the work. With transparent enamels the color beneath tends to intensify the beauty of the piece.

Steel, whether plated or not, is a poor thermal conductor; therefore, special care must be given to provide a uniform heat source when enameling silver-plated steel. Thus it becomes necessary to use a kiln for firing. The torch would tend to burn the silver in spots and cause it to flake off. Silver-plated steel needs no cleaning after it has been fired. There is no accumulation or firescale, and the work comes out of the kiln bright and clean.

Counterenameling is usually not necessary, except for the purpose of reinforcing the metal. The back as well as the front will have a beautiful finish, and will not tarnish.

Silver-plated steel is an excellent base for most enameling effects, including overlay and processes where the surface of the silver is not exposed. Controlled design and swirling are not suitable techniques to use on silver-plated steel.

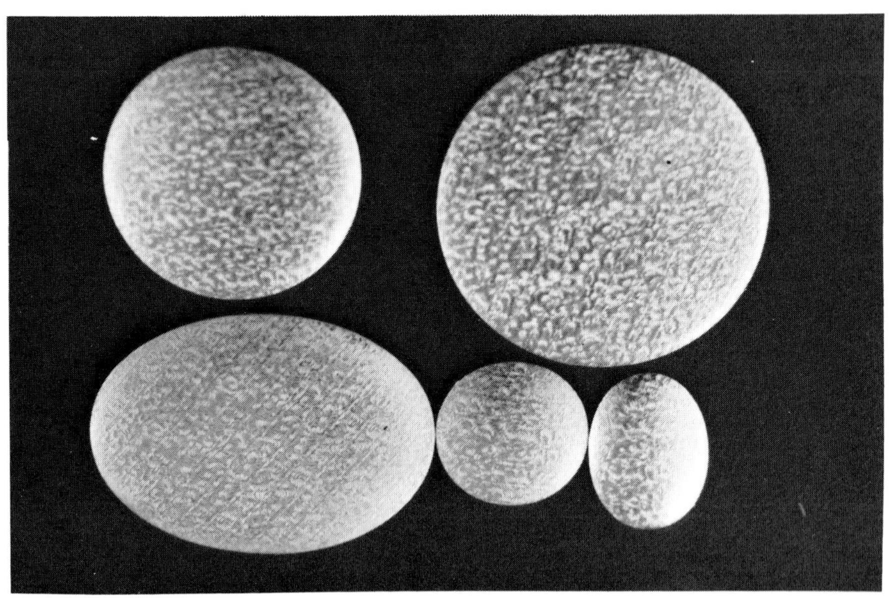

Examples of silver-plated, ready-shaped pieces

Procedures to Use When Enameling Silver-Plated Steel

A thin coat of binder, either 7001 enameling oil or Klyr-Fire, is applied to the spangled side of the piece. Number 450 clear flux is applied over the binder, in a uniform and smooth manner. This flux is made especially for this material. The flux used on copper will not produce satisfactory results. It is not sufficiently transparent to derive the fullest benefit from the silver metal beneath.

The work is fired in the kiln for about four minutes at about 1450°F. It is then removed from the kiln and allowed to cool. If a color is desired in addition to the color that the flux provides, a transparent enamel should be sifted directly over the binder. This can be accomplished with or without the flux.

Silver-plated steel with flowers

After the firing is complete and the work cooled, findings may be attached, and the piece is ready to use.

For decorative purposes, small lumps or threads can be utilized, or a design can be made with stencils.

CHAPTER SIXTEEN

Making Circular or "O" Ring Threads

An exciting variation in making threads is producing circular threads, or "O" rings. They are made by spiralling molten enamel around a special tool. The exciting aspect of the work is that the design possibilities are enormous. The rings may be cut into individual halves or quarters and placed in hundreds of ways, to form simple or intricate designs.

The size of the rings depends upon the size of the tool that is used in their construction. The tool is composed of three kinds of metals: copper, stainless steel, and mild steel. The variety of metals facilitates the handling by being good thermo-conductors and poor conductors. The tool is called a spiralling tool. Consult the diagram on page 94 in constructing one for use.

Instructions for Making a Spiralling Tool

Prerequisite to the actual making of circular threads is making the required tool. Since there are no such tools available on the market, it is necessary for the craftsman to make his own. The following are step-by-step instructions for making the spiralling tool:

1. Select a 3½-inch length of 1/2-inch copper pipe; this will serve as the barrel. Copper pipes are measured by their inside diameters; therefore, the rings made on this par-

ticular pipe will be slightly larger than one-half inch in diameter. After the student has acquired some experience, he may use different sized pipes, or tubing. For the first trial, 1/2-inch pipe is recommended.

2. File the ends to remove any burrs or rough places left from the cutting operation.
3. Using a hacksaw, cut six or eight V-shaped wedges into one end of the barrel. This makes it possible to close the end of the barrel around the handle.
4. The reason for using copper for the barrel is that copper is an excellent thermo-conductor and the heat that is generated by the hot enamel is dissipated. This will prevent the enamel spiral from sticking to the barrel.

Procedure for Constructing the Handle

1. The handle is made of ¼-inch mild-steel rod. It is possible to obtain this material at most hardware stores or welding shops. It is called mild steel welding rod, and often comes copper coated to prevent rusting. It is not an expensive material and is available in 36-inch lengths.
2. With a hacksaw, cut off one 9-inch length. Round off one end by filing or grinding it; this will improve the holding qualities of the handle. The other end will be hard-soldered to one end of the barrel. Steel is used because it is a relatively poor thermo-conductor. As a result, the fingers will be protected from the heat.

Procedure for Making the Probe

The third component of the spiralling tool is the probe. Almost any kind of mild steel may be used for this purpose. Stainless steel is the most desirable, but sometimes difficult. It is suggested that a common six-penny nail be used, as it is already pointed, and with the head cut off it will suffice. It should be cut to about ¾ inch in length.

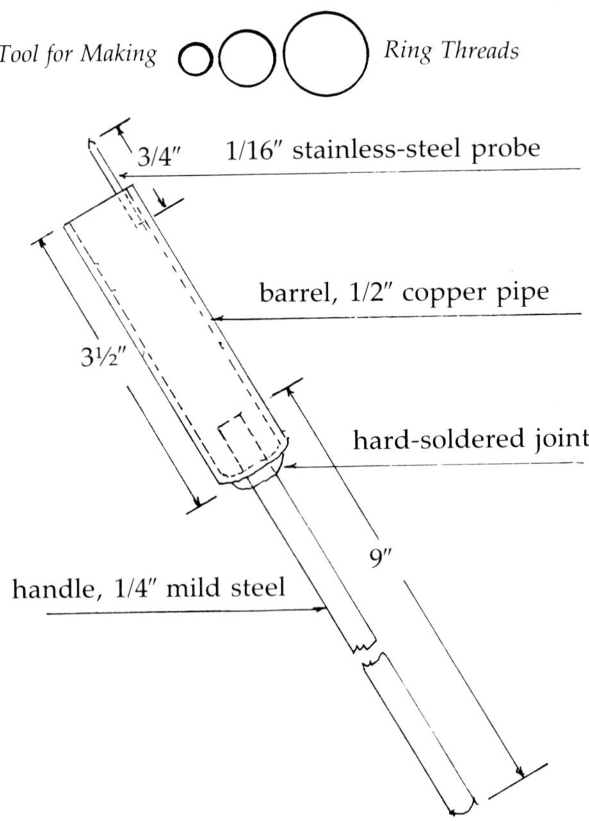

Procedure for Assembling the Spiralling Tool

It is a simple task to assemble the three parts. Consult the diagram above.

Step-by-Step Instructions for Making "O" Ring Threads

1. Place a copper cup on the grid, as described in the chapter on thread drawing. Fill the cup with the desired color of enamel, either opaque or transparent.

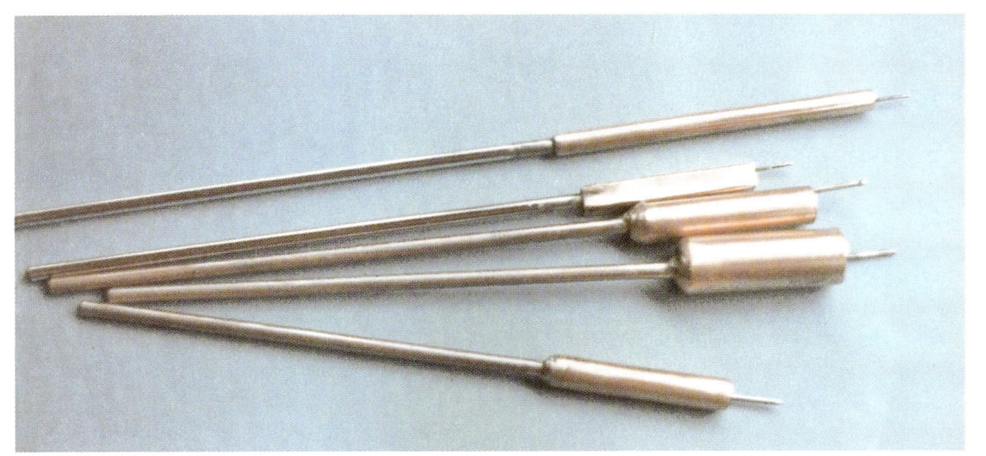

"O" ring tools

Examples of "O" ring usage

2. Heat the enamel in the cup with the torch, from the underside. When the enamel is completely melted and fluid, dip the probe into the enamel and hold it there for a second or two. When the enamel sticks to the probe, lift up the thread and wrap it around the barrel.
3. After the thread begins to take shape around the barrel, rotate the tool continuously, in one direction, and form a coil in a spring-like configuration.
4. The tool should be held rather close to the cup in order to prevent the enamel thread from becoming too cold. The greater the distance from the cup the tool is held, the cooler the thread, and consequently the looser it will be on the barrel. This could result in an uneven coil of threads.
5. When the coiled thread is complete and it is to be removed from the tool, use pliers to crack the portion that is on the probe. The whole coil should now slip off without breaking.

CHAPTER SEVENTEEN

Enameled Miniature Flowers

Miniature flowers are simple to fabricate and provide an enterprising adventure for the beginner. A bouquet of these little beauties, whether made of one color or mixed colors, makes a very attractive and appealing gift, or a ready conversation piece.

Tools and Materials Required for Making Miniature Flowers

Tools	Materials
Template (provided)	36-gauge copper foil or 20-gauge sheet
Coin or compass	Sparex No. 2 pickling solution
Scriber	No. 621A undercoat enamel for transparents
Scissors	Assorted transparent enamels
Tin snips	Assorted opaque enamels
Ball-peen hammer	Fine-line-black solution
Torch or kiln	Fine enamel threads, several colors
Firing grid	Small enamel lumps, assorted colors
Pliers	Nocorode soldering flux
Clamp or hemostat	Small pieces of 50/50 solder
Small brush, for flux	Sheet of clean paper, 8½″ x 11″
Block of wood with sump	7001 enameling oil
	No. "00" steel wool

Step-by-Step Instructions for Making Miniature Flowers

Forming the Flowers

1. Use either 36-gauge copper foil or 20-gauge copper sheet.
2. Both sides of the copper should be cleaned with "00" steel wool.
3. Decide on the size of the flower; 1 inch or 1½ inches is best.
4. Scribe a circle on the copper by using a coin or compass.
5. If using 36-gauge foil, cut out the circle with scissors; if 20-gauge copper is to be used, cut out the circle with tin snips.
6. Lay the copper circle on the paper template and mark the numbered divisions.
7. Using snips or scissors, cut out the sections that form the petals.
8. Form the flower by sinking the circle into a shallow dish lying on a block of wood that has a small hollowed recess. Use a ball-peen hammer to do the sinking.

Enameling the Flowers

9. Place a flower on a clean sheet of paper. Coat the concave side with 7001 enameling oil.
10. Carefully dust on the chosen color of enamel. Use No. 621A enamel undercoat under transparent enamel.
11. Add a few tiny lumps of a contrasting color to form the stigmas.
12. Fine threads or black-line solution may be used for anthers.

Firing the Flower

13. Place the flower in a level position on the grid.
14. If using a torch, heat the work from the underside; otherwise use a kiln.
15. When the enamel becomes completely fused and appears to be smooth and glassy, remove from the heat and cool.

Arrangement of miniature flowers

Miniature flowers combined with birds

Attaching the Stem

16. Flatten ¼ inch of one end of a 4-inch piece of 18-gauge copper wire; 20-gauge will also suffice.
17. Bend ⅛ inch of one end to a 90-degree angle to form the base of the stem.
18. Reshape the flattened portion to fit the back side of the flower.
19. Put a small amount of soldering flux on the end of the stem and on the back of the flower.
20. Place a small piece of 50/50 solder between the two pieces, sandwich style.
21. Use clamps or hemostat to secure the stem in place.
22. Using a torch, apply heat until the solder fuses. The flux will first boil to indicate readiness.
23. Allow the flower to cool, then clean by immersing in the pickling bath. Finish cleaning with steel wool.
24. The flower is now ready to exhibit.

Suggested Shapes and Forms for Making Miniature Flowers

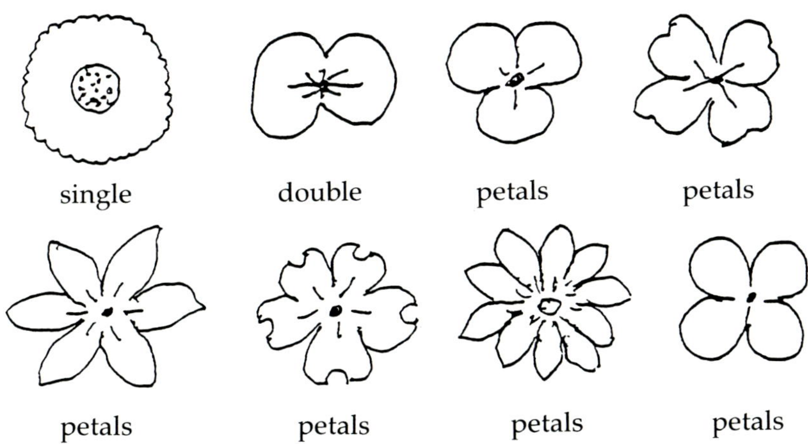

Suggested Shapes for Leaves

Templates for Marking Divisions

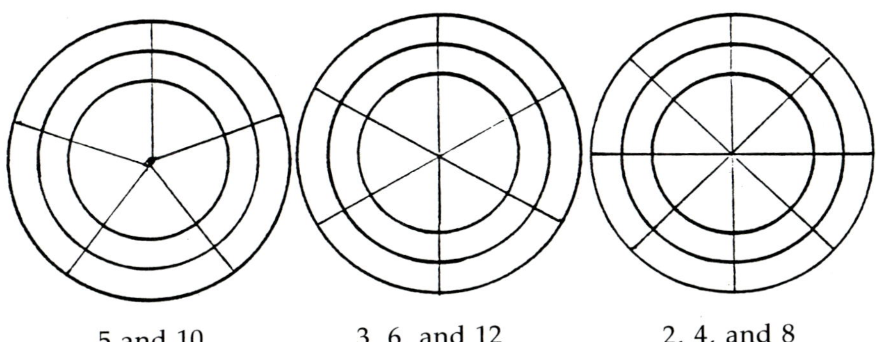

CHAPTER EIGHTEEN

Glass Beads

A recently developed process has made it possible for the hobbyist to make glass beads by using enamels. Pieces of red-hot copper tubing are rolled in enamel. The hot tubing picks up the enamel, which is then fused to it.

Necessary Materials

Copper tubing that is 1/8″, 3/16″, or 1/4″ or larger in diameter is used for the core of the bead. The tubing may be purchased in 50-foot coils or in 1-foot pieces and cut into suitable lengths for beads.

The following types of enamels are required:

1. Soft-fusing clear flux powder
2. White particles in 8/14 mesh
3. Several selected colors of enamel in 80 mesh
4. Several selected colors of threads and lumps

Necessary Equipment

Two pieces of insulation board, each approximately 6″ x 8″, are needed. One board is used as a place for collecting and cooling the beads. The other one is used in making beads.

A torch is essential, and the acetylene torch is the ideal type for bead making; however, a propane or butane torch will suffice. The torch should be wired or clamped to the bench or grid as a safety measure.

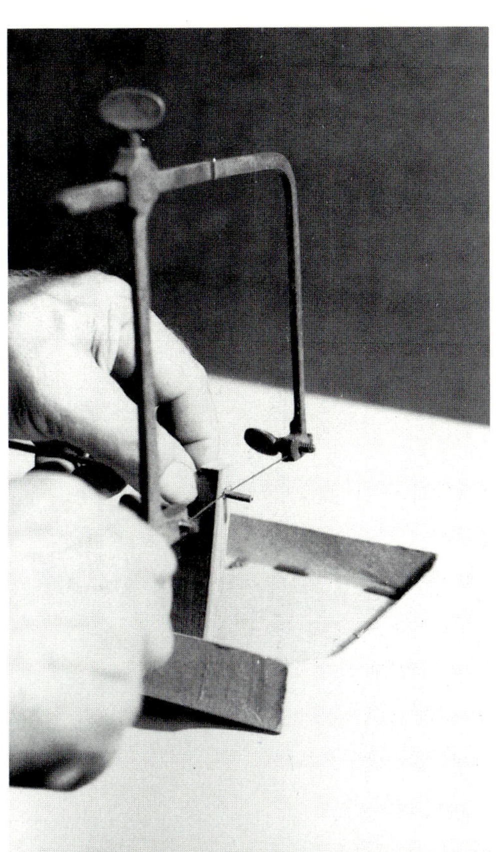

Device for measuring and sawing copper tubing

Jig for holding copper tubing while sawing off lengths

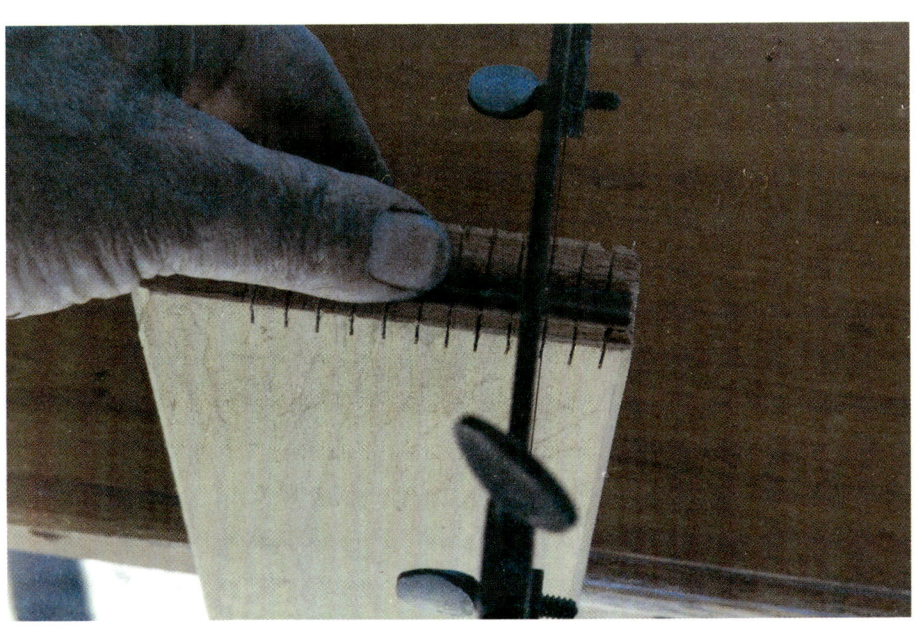

A device is needed in order to cut the tubing into sizes suitable for beads. This device is made in the following way:

1. A block of wood approximately 1″ x 4″ x 4″ is securely fastened to the bench, the end-grained edge being uppermost. It can be fastened by using a vise, screws, or clamps.
2. A "V" groove is sawed in the uppermost edge of the block. The groove should extend the full length of the block and be placed midway between the two sides. The depth of the groove should be not less than the diameter of the tubing.
3. Saw-cuts at 1/4-inch intervals are made across the thickness of the block. They extend to the bottom of the groove and serve as a guide in measuring and cutting the tubing.
4. The tubing is placed in the groove, with one end flush with the end of the block.
5. At a selected saw-cut the tubing is cut with a jeweler's saw. A five-inch "00" blade is used. The tubing is then pushed over until it is again even with the edge of the block, and another piece is cut in the same manner.

Necessary Tools

A firing rod is needed, and it is made by using a nine-inch length of stainless-steel welding rod. The diameter of the rod should be slightly larger than the inside diameter of the tubing. About half an inch of the rod is made smaller by filing it to a diameter that allows it to fit into the tubing. A firing rod made by this method has a shoulder against which the tubing rests when it is being enameled.

A tool is needed to remove the hot finished bead from the firing rod. A knife or spatula can be used.

A six-inch mill file is needed to remove any exposed copper extending beyond the glass portion of the finished bead.

Materials needed to begin bead making: copper tubing, firing rod, various enamels

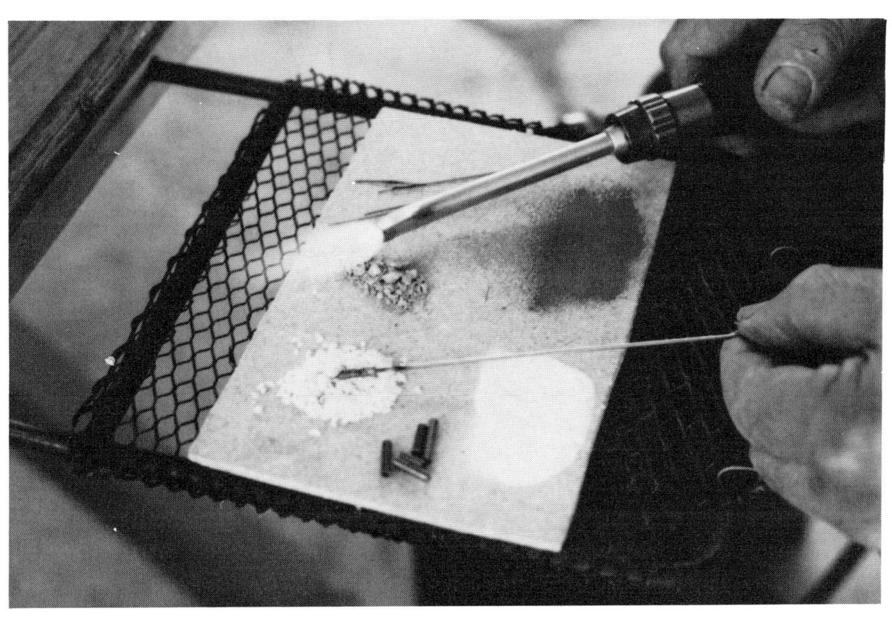

Picking up small particles of enamel for building up the bead

Procedure for Making Glass Beads

1. Several bead-length pieces of tubing are cut in the tube-cutting device.
2. A small amount of soft flux powder, a few grains of 8/14 mesh white particles, and a small pile of the selected color of 80-mesh powder are placed on one of the asbestos boards.
3. The powders are flattened with a spatula to a thickness of about 1/32 inch.
4. A piece of the tubing is placed against the shoulder on the firing rod.
5. The torch is lighted and the tubing is heated until it is red hot.
6. The hot tubing is rolled in the flux until the entire surface is coated.
7. The tubing is again heated in order to melt the flux that has been gathered up on it.
8. Now the hot tubing is rolled in 8/14 mesh, picking up white particles. Care must be exercised to prevent the enamel from being fused to the rod.

   The tubing is again heated and must be continuously rotated while the particles are being fused in order to maintain a balanced effect. At this stage the bead begins to take on the final shape and size.
9. Step 8 is repeated as many times as necessary or until the bead is approximately the finished size and shape.
10. Now the bead is reheated, rolled in the selected color of 80-mesh powder, and refired again. This operation must be repeated three or four times in order to produce a bead that is well coated and symmetrical.
11. As a final measure, the bead may be heated again and threads or lumps added for special effects. The bead is now fired for the last time; it is rotated continuously to insure a uniformity of shape and temperature.
12. The bead is allowed to cool slightly. It is then removed from the firing rod and placed on the insulation material for further cooling. If it is permitted to remain on the rod for too long a period, it is likely to stick to the rod and be difficult to remove.

13. Any exposed portion of the copper tubing should be filed away at this time. The bead is now finished and ready for stringing.

Notes Pertaining to Bead Making

1. The length of the bead is determined by the length of the tubing, and the diameter of the bead by the amount of build-up.
2. The shape of the bead is determined by the angle at which it is held in the flame. Slanting the rod will cause the bead to be bulbous on one end. If the rod is held in a horizontal position, a more uniform shape will result.
3. Uniform heating is assured by continuously rotating it while it is in the flame. If one side becomes too hot, the enamel will sag and the bead will become distorted. A bead that has been unevenly fired may have a tendency to crack or shatter.
4. Sometimes the bead becomes discolored because of excess carbon in the flame. If this occurs, the bead should be moved to the outer portion of the flame, where it is farther away from the torch tip. The outer portion of the flame burns hydrogen from the tank and oxygen from the air, and so there is less chance for the bead to become carbonized or stained. Staining is a greater problem when transparent enamels are used. In most cases, the stain will disappear with additional heating.
5. Transparent enamel powders used over opaque white produce especially pleasing effects.

Instructions for Making a Bracelet by Using Enameled Beads

1. The core of the bead requires 3/16-inch copper tubing.
2. Twenty-two pieces of tubing are cut to 3/4-inch in length.
3. The pieces of tubing are coated with flux and three coats of selected enamel. Extra build-up is not needed; therefore, 8/14 mesh particles are not used. The beads are oth-

Removing the bead from the rod

A string of finished beads

More finished beads

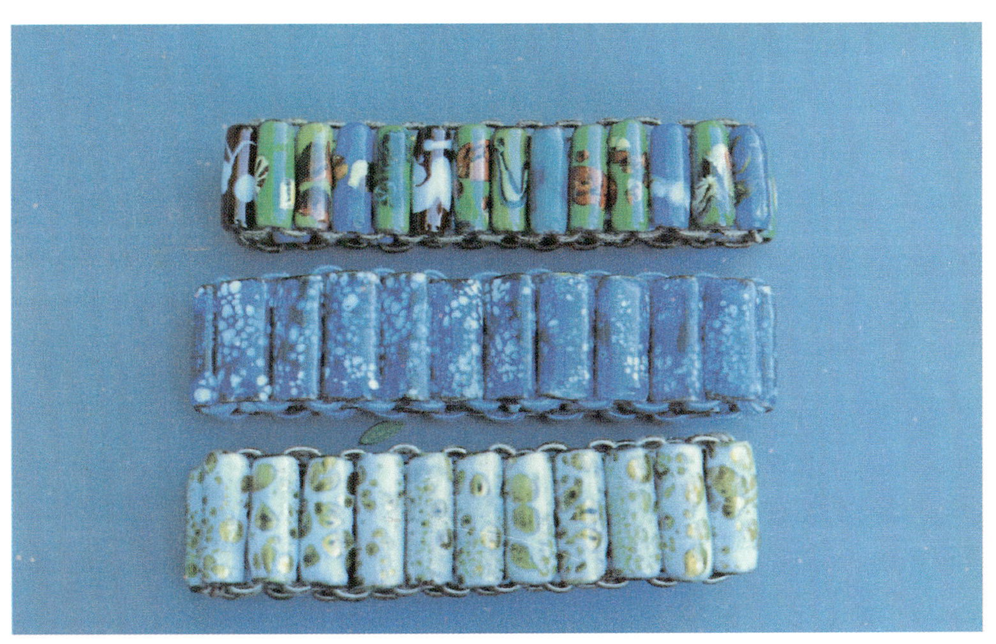

Use of beads to make bracelet

erwise made according to the previously described procedure.

4. When the twenty-two beads are finished, they are strung together using 1¼ yards of 1/8-inch round elastic. The elastic can be purchased at most notion stores and tinted to match the color of the beads.
5. The first bead to be strung is placed midway between the two ends of the elastic. The second bead is placed parallel to the first and as close to it as possible. The two ends of the elastic are held firmly and passed through the second bead from opposite ends. The third bead is strung in the same manner, and the procedure is continued until all of the beads are used. At all times twisting of the elastic must be avoided.
6. When the beads have all been strung the two ends of the elastic are pushed through the first bead. The circle of beads is now complete and the elastic is tightened.
7. The two ends of the elastic now protrude from opposite ends of the first bead. The pieces of elastic are stretched tightly and a knot is tied in each as close to the bead as possible. The excess elastic is cut off and the knots are tucked into the bead.

Bead-Making Technique as Applied to Bolo Tie Tips

Preliminary Requirements

1. A small piece of 20-gauge sheet copper.
2. A metal template or paper pattern made according to the accompanying illustration.

Procedure

1. The template or pattern is placed on the sheet of copper and traced with a sharp pencil or stylus.
2. The copper shape is cut out with tin snips.

This little device is designed to assist the craftsman during the process of stringing the beads. It is made of 90-degree aluminum angle, which is cut to a length of 18 inches with 3/4-inch sides. The two pieces of the angle are cut to a length of one inch and riveted to the center of the main body. These two short pieces support the device and aid in preventing it from rolling over. The main body of the angle is marked off at 3-inch intervals to help in determining the length of the strand. The beads are placed at random along the channel and sorted to suit, regarding shapes, colors, and so forth. After all arrangements have been completed, the beads may be strung without being removed from the device. Tiger Tail is a splendid material to use for this purpose.

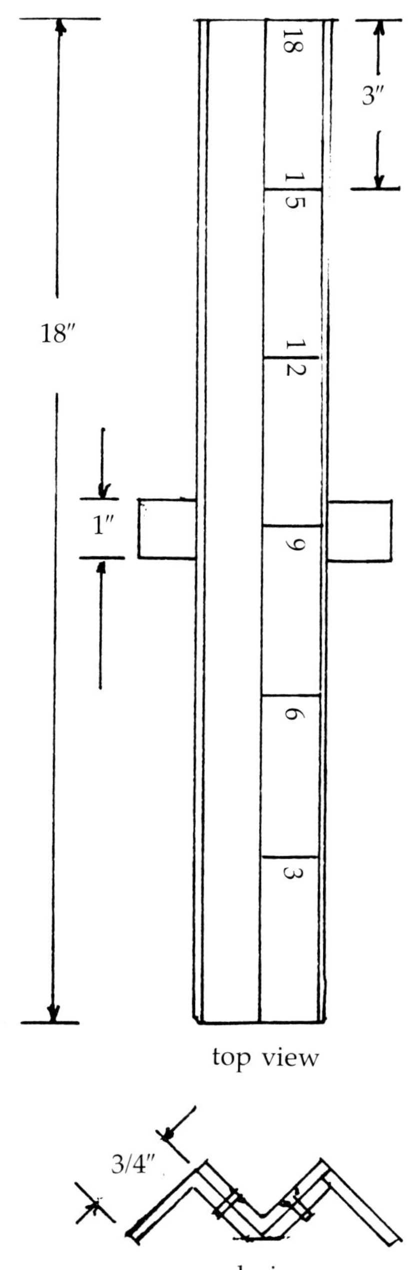

(Scale, 1″ = 3″)

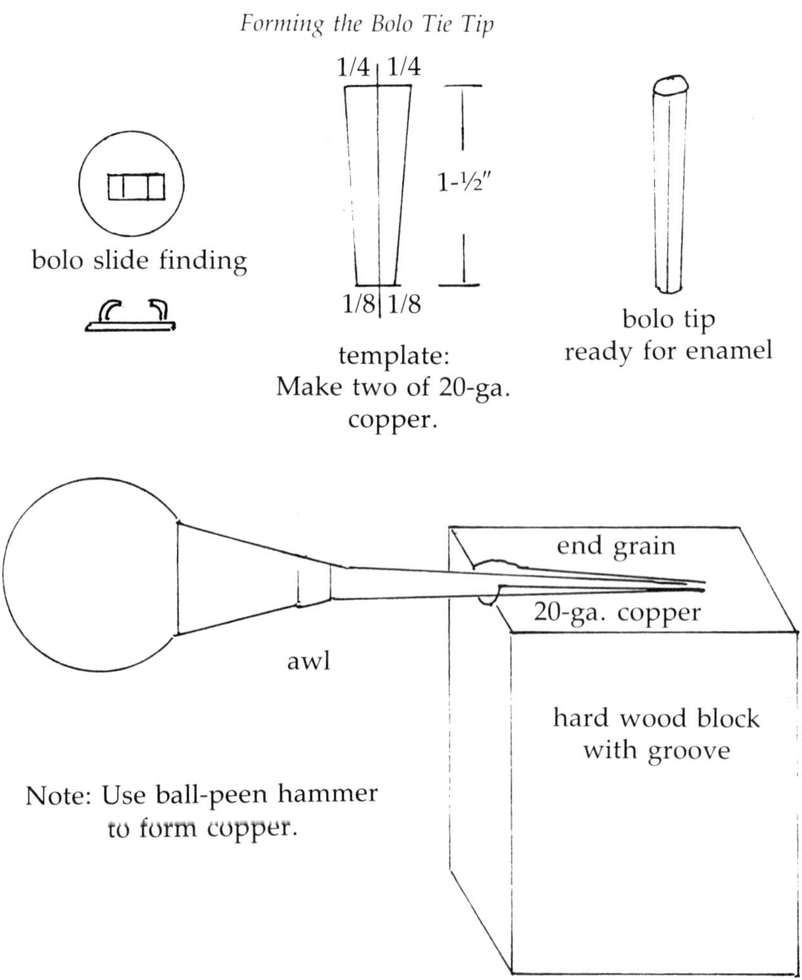

3. The shape is formed into the core for the bolo tip by hammering it around a tapered piece of steel, such as an awl. The two long edges should be joined without overlapping.
4. The core is placed on the firing rod and enameled in accordance with the directions given for the bracelet bead, one coat of flux and three coats of a chosen powdered enamel being sufficient.
5. As a final touch, the small end can be neatly closed by dipping it into the enamel powder and fusing it.

6. The bolo tip is now finished and ready for removal from the firing rod.

Completing the Bolo Tie

A 36-inch cord, a decorative enameled tie slide, two bolo tips, and an appropriate finding are necessary. The cord and finding can be purchased and the enameled slide designed and made by the craftsman. The finding is soft-soldered to the back of the enameled slide.

Variety of bolo ties showing tips made as explained in text

CHAPTER NINETEEN

Floral Wafers

Working with floral wafers offers exciting possibilities, and this is a natural follow-up to thread drawing.

Small wafers of enamel that are made to be used as embellishments are known as floral wafers. They are placed on prefired enameled surfaces or on unfired surfaces that have been dusted with enamel powder. The undercoat and the floral wafers are then fired in one operation by using either the kiln or the torch.

Essential Techniques

Floral wafers are made by arranging enamel powder, lumps, and threads in a copper cup like the one described in the chapter on thread drawing. The enamel in the cup is fused by using a torch and a thread is drawn according to the instructions given in the lesson on thread drawing.

After the thread is cool, it is cut into small sections, which are the finished floral wafers, the thickness of these wafers being about one-third of their diameter.

Composition

It is necessary to understand what actually happens when two or more colors are simultaneously fused and a thread is drawn. The color that is placed in the bottom of the cup becomes the center of the finished wafer. The next color to be placed in the cup surrounds the first color. Any additional colors will, ac-

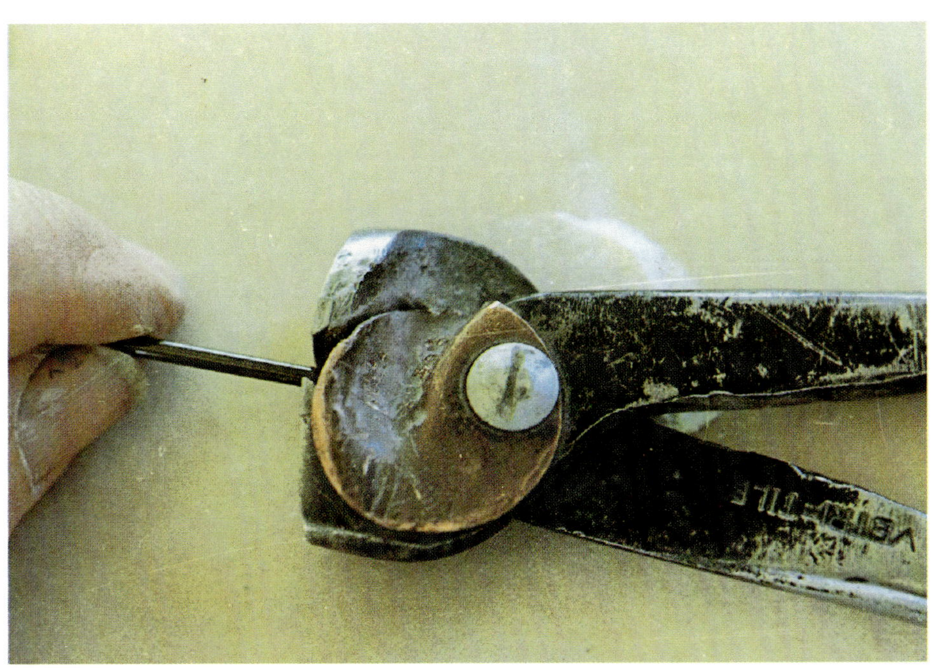

Nippers with apron, used for cutting threads

cording to their position in the sequence, be closer and closer to the outside rim of the wafer. If the wafer, for instance, is to be red with a white center, the white will be used to fill the bottom half and the red the remaining half. The thread produced from this combination will be red but will have a white center running parallel to its length. Floral wafers cut from this thread will, as a rule, be red with white centers. Should the cup be filled with half blue and half green, each color occupying one side of the cup, the thread will have a green side and a blue side running the full length of it and the wafers will be half green and blue.

Floral-Wafer Cutting

The wafers are cut with tile-worker's nippers. Here a problem presents itself because the wafers have a tendency to scatter in all directions when being cut. This problem is solved by employing either of the following measures.

Transparent Bag Method

A small transparent plastic bag is required. One end of the thread is pushed into the bag through a small hole in the closed end. The thread remaining outside of the bag is held firmly with one hand. The other hand is free to manipulate the nippers. The nippers are held inside the bag, where the wafers are cut and accumulated. It is possible to see through the plastic to determine whether or not the wafers are being cut properly.

Nippers with Added Aprons

This is the sophisticated method, and it should be resorted to when a considerable amount of cutting is necessary. An improvised apron or guard is attached to each side of the nippers to prevent the particles from escaping. These aprons are about one inch in diameter, made of copper, and shaped to fit the jaws of the nippers. They are fastened to the nippers with small bolts or matching screws.

Methods Used in Making Special Floral Wafers

The most outstanding wafers are those that portray flowers. Instructions for making a variety of flower wafers follow. All of these are made by using medium fusing enamel. The student should not confine his efforts to these specific examples. He will discover by experimentation that he can create many beautiful and original wafers.

A Fundamental Type (Sunflower Used on Any Background)

1. About half a teaspoon of opaque brown enamel is placed in the bottom of the cup and leveled and packed with the spoon.

2. The cup is then filled to within 1/8 inch of the top with opaque yellow enamel and is leveled and packed.
3. The next step requires the use of threads, these having been made previously, of the same brown that is used in the bottom of the cup. There should be twelve to sixteen pieces of thread, each approximately 1/32 inch in diameter and 3/8 inch in length.

   Tweezers are used to place the threads on the yellow enamel. They are arranged in a radiating pattern, touching each other near the center of the cup and pointing outward toward the rim. These threads become the petals of the finished flower.
4. The enamel is now ready to be fused and the thread to be drawn.

Conventional White Flower (To Be Used on a Blue Background)

1. The cup is filled and packed with opaque blue enamel to within 1/8 inch of the top.
2. A hole or well is made in the center by pushing the unsharpened end of a pencil through the enamel to the bottom of the cup. This is done cautiously in order to prevent the enamel from caving back into the hole.
3. The hole is then filled with opaque enamel. This enamel will be the center of the flower. In order to fill the hole easily a small funnel is used.
4. Five to eight opaque lumps about the size of small peas are selected. They are placed on the surface of the blue enamel in a uniform pattern midway between the center and the rim.
5. The enamel is now ready for fusing and thread drawing.

In order to obtain the most desirable effect, the finished wafers are used on a surface that is enameled with the same blue enamel that is used in Step 1. The blue in the wafer is identical to the blue in the background and will not show in the finished wafer.

Piece showing use of floral wafers

One method of preparing floral wafers

Conventional White Flower
(To Be Used on Contrasting Background)

When these wafers are fired on any color other than the opaque blue, the effect will be quite different. The opaque blue in the wafer will appear as a matrix surrounding the petals and the center. The flower may not be as outstanding as the one fired on the opaque blue, but it will be satisfactory. By using different backgrounds any number of effects can be obtained.

A Simple White Daisy Type

1. The cup is packed to the top with opaque Chinese red enamel.
2. Eight white threads 1/32 inch in diameter and 3/8 inch in length are placed evenly on the red enamel, touching at the center and radiating to the rim of the cup.
3. The cup of enamel is fired and threads are drawn.

These are easy to make and are very dependable. They are at their best when used on a Chinese red background, but they can be used successfully on almost any background.

The Cutting-In System
(Using Powdered Enamel)

1. The cup is filled to within 1/8 inch of the top with three different colors of opaque enamel: yellow, white, and orange. White occupies one-fourth of the space, yellow occupies one-half (one-fourth adjoining each side of the white), and the remaining one-fourth is filled in with orange. The two yellows are opposite and the white and orange are opposite.
2. The enamel is firmly packed and then six grooves or wedges are made in it with the palette knife. The grooves extend to the bottom of the cup and are about 1/8 inch

Another method of making floral wafers

Use of daisy flowers in tray

The cutting-in method

wide at the rim, tapering off as they reach the center.

3. The grooves are filled with medium fusing flux. They must be made and filled one at a time in order to prevent the walls from caving in. Each groove is made and filled with flux before proceeding to the next groove.
4. Upon completion of the last groove, the enamel is firmed. A hole is made in the center of the enamel and filled with light brown enamel.
5. The enamel is then fired and the wafer thread is drawn.

These floral wafers can be used on any contrasting background; transparent green is suggested.

Floral Wafers Used as Petals of Flowers

Floral wafers can be made especially to be used as single flower petals. They are a delightful variation of the usual wafer and are very easy to make.

Cutting method for making single petals

Samples of wafers

1. The cup is filled to within 1/8 inch of the top with any color of enamel, and the enamel is packed down and leveled.
2. With the palette knife, a single groove or wedge is made in the enamel from the rim of the cup to the center. The wedge extends to the bottom of the cup and is about 1/8 inch wide at the rim, the sides of the wedge coming together in the center.
3. The wedge is filled with either clear flux, or opaque or transparent enamel.
4. The enamel is fused, a thread drawn, and the wafer cut.

   Floral wafers made by this method are used as individual petals, and five or more are assembled to make a flower. They are arranged as petals, the large ends of the wedges being placed near the center of the flower.
5. A lump or cluster of small fragments of some suitable color can be used as the center of the flower.

Floral Wafers Used as Leaves

Floral wafers can be made especially to represent leaves. The method used in making leaves is the same as that used in making petals. The flux or powdered enamel that is placed in the wedges becomes the veins of the leaves.

Applying the Leaves

1. The leaves are positioned on the enameled piece with the wedge sides as the bases of the leaves.
2. The piece is heated with the torch until the enamel is completely fused.
3. The copper-tipped swirling rod is used to pull the ends of the leaves out slightly to make them appear pointed.
4. The bases of the leaves are pulled inward to give them a more leaflike appearance.

A piece showing single petals and leaves

An example of the use of special floral wafers

A variety of wafers

A variety of single petals arranged on a tray

CHAPTER TWENTY

Related Metal-Working Processes

This chapter is devoted to those processes that are related to enameling but do not necessarily involve the actual use of enamels.

Most of the methods of working with metals, however, can include enameling with satisfactory results, provided that special attention is given to design and colors.

This chapter is intended to familiarize the beginner with the aspects of metal working that relate to enameling. The craftsman should, at least in a small measure, be acquainted with those techniques mentioned here.

They are as follows: etching, copper tooling, chasing, metal sculpture, and pileation. These are the processes used to create artistic pieces such as trays, plaques, coasters, jewelry, and other forms of pleasing work.

Etching

The term *etching* implies that metal is being removed, or dissolved, by the chemical action of acids. This factor alone behooves one to exercise extreme caution, because of the dangers involved. It is important to guard against spilling the acid or breathing the fumes. One must be especially careful in diluting the acids. Always dilute acids by adding acid to water, not water to the acid. If the latter is done, the reaction may be violent and result in severe burns. When acids are used for etching, they are called

Etched copper, no enamel

Etched copper tray with embossed edges, not enameled

mordants. On copper, brass, and bronze, the proper acid to use for etching is nitric acid (HNO_3). For steel, aluminum, and tin alloys, the proper acid is diluted hydrochloric acid (HCl).

When preparing to etch copper, it is important to select a piece of metal that is the correct size, shape, and gauge.

The section of the design that is to remain unetched and in relief is painted with a petroleum product called ground or resist. It is applied with a brush to the places where the design is to be saved. The design space should be full of design to avoid etching away more metal than is necessary.

Etching is an art form and must be treated as such. Many beautiful artifacts and wonderful pieces can be the result of careful etching.

Sometimes a design is produced in which a series of cells or basins is etched out of the metal. These cells are filled with the appropriate color and quantity of enamel, and fired. It may be necessary to fill and fire several times to bring the enamel level with the metal surface.

The French have a name for this procedure and use of enamel. They call it *champleve.* An example of champleve is shown in the picture of the bolo tie.

Copper Tooling

In an annealed condition, copper is a very malleable and pliable metal. It can be worked, stretched, or drawn into intricate patterns and shapes.

Tooling is the term used to describe one of the various techniques in which copper is forced into deep cavities, or high relief.

For this method of working with copper, thirty-six–gauge is most often used. It is listed in trade catalogs as tooling copper and is usually already annealed.

The thin copper is worked from the back side with special tools. First, the design is drawn on the face side. It is then outlined with a chasing tool or stylus. The work is then turned over, back side up, and, with care for the design, pressed into shape. A mass of glazier's putty is placed under the work to act as a holding agent and cushion at the same time.

Copper tooling, no enamel

Care must be exercised to prevent the tools from breaking through the copper. When the work is completed and everything is satisfactory, the piece is reinforced by adding some sort of material to the underside. On small pieces it is seldom necessary to reinforce the work. On large toolings, 50/50 solder can be used with success, and is an excellent choice.

Chasing

Chasing, as it applies to the ornamentation of metals, is accomplished with a set of small tools and a chasing hammer. First, the design is traced or drawn on the metal. The metal must be supported on something solid to assist the tool in doing its work. With the tool held at a slight angle, and tapped with a hammer,

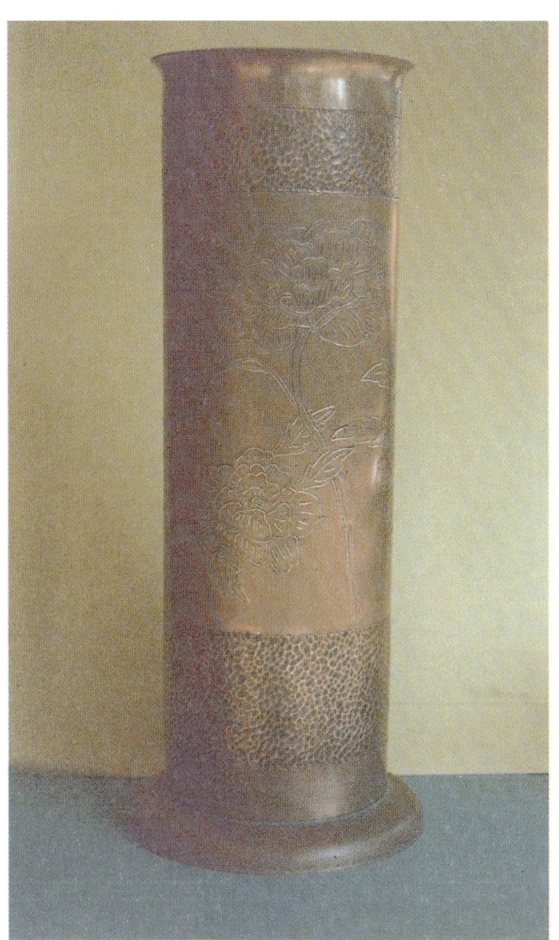

Umbrella stand, chased design. No enamel.

Aluminum tray with chased design. No enamel used.

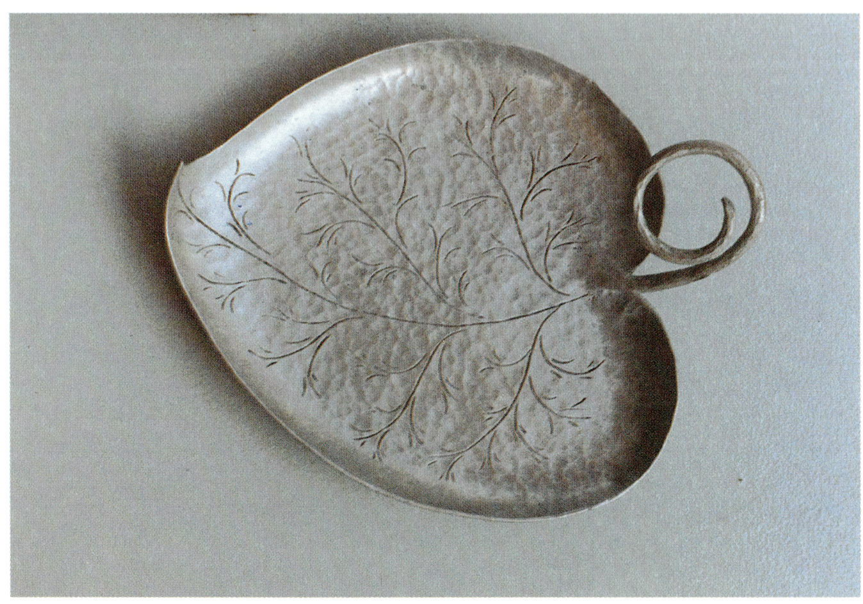

it will tend to creep forward, causing small marks in the metal surface. As the work continues, the tool will produce a continuous line. The marks become joined, and a chased line is the result.

This technique is often applied to silver pieces, to embellish and enhance their beauty.

The umbrella stand shown in the photo on page 130 was made of copper pipe. It was hand-chased by the author. After the chasing was complete, the base was soldered to the bottom.

Metal Sculpture

Metal sculpture is one of the techniques where enameling may be introduced. The term *sculpture* implies that the work be in the three dimensions and be balanced in design and contour.

Metal sculpturing is an ideal process for making wall hangings, framed or unframed. Almost any decorative creation may be utilized, such as flowers, birds, leaves, or butterflies. Almost any motif or original idea can be prepared for enameling. The enamels add to the colorful and pleasurable effects.

Fine colors may be brought out by flame heating instead of enameling. The oxides that appear when copper is heated carefully are extremely brilliant, and not too difficult to obtain. The colors range from the original copper color through a golden yellow to a light green, dark green, several shades of blue, lavender, several shades of red, and many in-between shades. The different colors appear in rapid succession, and can only be controlled by slow and careful temperature regulation. A small oxyacetylene torch is the best heating tool for this purpose. A glance at the diagram on page 132 will offer some assistance in understanding the chemistry of the flame.

The torch is supplied with acetylene gas (C_2H_2) and oxygen (O_2). When the torch valves are opened and an equal amount of each gas is supplied, the flame is known as a "neutral flame." Any excess of acetylene will cause the flame to produce smoke and will deposit soot on the copper. This condition is known as carbonizing, and it is not desirable. When the flame contains an excess of oxygen, it is referred to as an oxidizing flame. It is this flame that produces the most brilliant colors on copper. If the

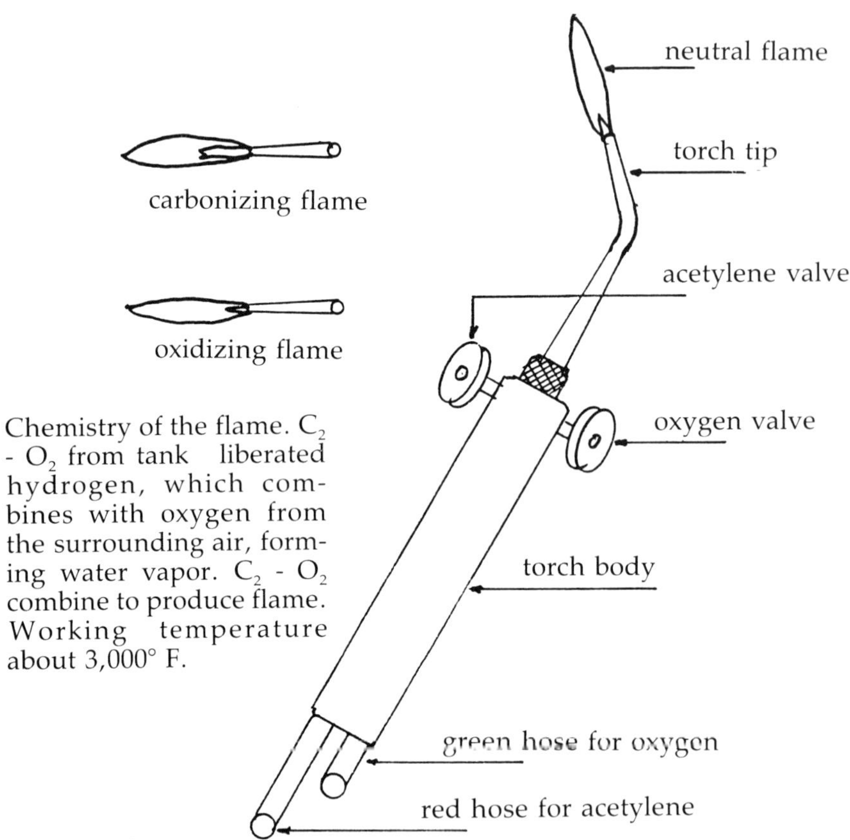

flame is played directly on the copper and careful control of the temperature is maintained, the oxygen will react with the copper and cause a film of copper oxide to be deposited on the surface of the work.

These colors are beautiful and very pleasing, and they will endure a long time. There is no need to use a protective coating such as wax or lacquer. These materials tend to dull the final finish and are not at all necessary.

Pileation

A small oxyacetylene torch is a very useful tool in decorating the edges of trays, coasters, and many other small items. The torch flame is adjusted for a neutral flame. The edge of the piece is melted to form beads of molten metal along the circumference of the piece.

The procedure has a name: the metal is being *pileated*. It tends to reinforce the article as well as decorate it.

The article that is to be pileated is heated all over to a red heat in order to anneal it. This makes it possible for the craftsman to continue with the edge work. The flame should be held very close to the metal and moved slowly around the piece. Care must be exercised to prevent the edge from becoming too hot. If this happens, the metal will sag and possibly destroy the design or work.

Metal sculpture colored with heat. No enamel added.

Metal sculpture of copper with everything enameled

*Metal sculpture.
Everything is enameled.
Notice pileated edges.*

Hand-wrought roses.
No enamels used.

Hand-wrought flowers no enamel.